Springer Undergraduate Mathematics Series

Springer
London
Berlin
Heidelberg
New York
Barcelona
Budapest
Hong Kong
Milan
Paris
Santa Clara
Singapore
Tokyo

Other books in this series

Basic Linear Algebra *T.S. Blyth and E.F. Robertson (3-540-76122-5)*
Elements of Logic via Numbers and Sets *D.L. Johnson (3-540-76123-3)*
Multivariate Calculus and Geometry *S. Dineen (3-540-76176-4)*
Elementary Number Theory *G.A. Jones and J.M. Jones (3-540-76197-7)*
Vector Calculus *P.C. Matthews (3-540-76180-2)*
Introductory Mathematics: Algebra and Analysis *G. Smith (3-540-76178-0)*
Introductory Mathematics: Applications and Methods *G.S. Marshall (3-540-76179-9)*
Groups, Rings and Fields *D.A.R. Wallace (3-540-76177-2)*
Measure, Integral and Probability *M. Capinksi and E. Kopp (3-540-76260-4)*

Zdzisław Brzeźniak and Tomasz Zastawniak

Basic Stochastic Processes

A Course Through Exercises

With 21 Figures

Springer

Zdzisław Brzeźniak, PhD
Tomasz Zastawniak, PhD
Department of Mathematics, University of Hull, Cottingham Road, Hull HU6 7RX, UK

Cover illustration elements reproduced by kind permission of:
Aptech Systems, Inc., Publishers of the GAUSS Mathematical and Statistical System, 23804 S.E. Kent-Kangley Road, Maple Valley, WA 98038, USA. Tel: (206) 432 - 7855 Fax (206) 432 - 7832 email: info@aptech.com URL: www.aptech.com
American Statistical Association: Chance Vol 8 No 1, 1995 article by KS and KW Heiner 'Tree Rings of the Northern Shawangunks' page 32 fig 2
Springer-Verlag: Mathematica in Education and Research Vol 4 Issue 3 1995 article by Roman E Maeder, Beatrice Amrhein and Oliver Gloor 'Illustrated Mathematics: Visualization of Mathematical Objects' page 9 fig 11, originally published as a CD ROM 'Illustrated Mathematics' by TELOS: ISBN 0-387-14222-3, German edition by Birkhauser: ISBN 3-7643-5100-4.
Mathematica in Education and Research Vol 4 Issue 3 1995 article by Richard J Gaylord and Kazume Nishidate 'Traffic Engineering with Cellular Automata' page 35 fig 2. Mathematica in Education and Research Vol 5 Issue 2 1996 article by Michael Trott 'The Implicitization of a Trefoil Knot' page 14.
Mathematica in Education and Research Vol 5 Issue 2 1996 article by Lee de Cola 'Coins, Trees, Bars and Bells: Simulation of the Binomial Process page 19 fig 3. Mathematica in Education and Research Vol 5 Issue 2 1996 article by Richard Gaylord and Kazume Nishidate 'Contagious Spreading' page 33 fig 1. Mathematica in Education and Research Vol 5 Issue 2 1996 article by Joe Buhler and Stan Wagon 'Secrets of the Madelung Constant' page 50 fig 1.

ISBN 3-540-76175-6 Springer-Verlag Berlin Heidelberg New York

British Library Cataloguing in Publication Data
Brzeźniak, Z.
Basic stochastic processes : a course through exercises.
(Springer undergraduate mathematics series)
1. Stochastic processes - Problems, exercises, etc.
I. Title II. Zastawniak, Tomasz Jerzy
519.2'3'076
ISBN 3540761756

Library of Congress Cataloging-in-Publication Data
Brzeźniak, Zdzisław, 1958-
Basic stochastic processes : a course through exercises - Zdzisław Brzeźniak and Tomasz Zastawniak.
p. cm. -- (Springer undergraduate mathematics series)
ISBN 3-540-76175-6 (Berlin : pbk. : acid-free paper)
1. Stochastic processes I. Zastawniak, Tomasz, 1959-
II. Title. III. Series.
QA274.B78 1998 98-7021
519.2--dc21 CIP

Printed in Great Britain
2nd printing, with corrections, 1999

Typesetting: Camera ready by authors and Michael Mackey
Printed and bound at the Athenæum Press Ltd., Gateshead, Tyne & Wear
12/3830-54321 Printed on acid-free paper SPIN 10738037

To our families

Preface

This book has been designed for a final year undergraduate course in stochastic processes. It will also be suitable for mathematics undergraduates and others with interest in probability and stochastic processes, who wish to study on their own.

The main prerequisite is probability theory: probability measures, random variables, expectation, independence, conditional probability, and the laws of large numbers. The only other prerequisite is calculus. This covers limits, series, the notion of continuity, differentiation and the Riemann integral. Familiarity with the Lebesgue integral would be a bonus. A certain level of fundamental mathematical experience, such as elementary set theory, is assumed implicitly.

Throughout the book the exposition is interlaced with numerous exercises, which form an integral part of the course. Complete solutions are provided at the end of each chapter. Also, each exercise is accompanied by a hint to guide the reader in an informal manner. This feature will be particularly useful for self-study and may be of help in tutorials. It also presents a challenge for the lecturer to involve the students as active participants in the course.

A brief introduction to probability is presented in the first chapter. This is mainly to fix terminology and notation, and to provide a survey of the results which will be required later on. However, conditional expectation is treated in detail in the second chapter, including exercises designed to develop the necessary skills and intuition. The reader is strongly encouraged to work through them prior to embarking on the rest of this course. This is because conditional expectation is a key tool for stochastic processes, which often presents some difficulty to the beginner.

Chapter 3 is about martingales in discrete time. We study the basic properties, but also some more advanced ones like stopping times and the Optional Stopping Theorem. In Chapter 4 we continue with martingales by presenting

Doob's inequalities and convergence results. Chapter 5 is devoted to time-homogenous Markov chains with emphasis on their ergodic properties. Some important results are presented without proof, but with a lot of applications. However, Markov chains with a finite state space are treated in full detail. Chapter 6 deals with stochastic processes in continuous time. Much emphasis is put on two important examples, the Poisson and Wiener processes. Various properties of these are presented, including the behaviour of sample paths and the Doob maximal inequality. The last chapter is devoted to the Itô stochastic integral. This is carefully introduced and explained. We prove a stochastic version of the chain rule known as the Itô formula, and conclude with examples and the theory of stochastic differential equations.

It is a pleasure to thank Andrew Carroll for his careful reading of the final draft of this book. His many comments and suggestions have been invaluable to us. We are also indebted to our students who took the Stochastic Analysis course at the University of Hull. Their feedback was instrumental in our choice of the topics covered and in adjusting the level of exercises to make them challenging yet accessible enough to final year undergraduates.

Contents

1
Review of Probability

In this chapter we shall recall some basic notions and facts from probability theory. Here is a short list of what needs to be reviewed:

1) Probability spaces, σ-fields and measures;
2) Random variables and their distributions;
3) Expectation and variance;
4) The σ-field generated by a random variable;
5) Independence, conditional probability.

The reader is advised to consult a book on probability for more information.

1.1 Events and Probability

Definition 1.1

Let Ω be a non-empty set. A *σ-field* $\mathcal{F}$ on Ω is a family of subsets of Ω such that

1) the empty set $\emptyset$ belongs to $\mathcal{F}$;
2) if A belongs to $\mathcal{F}$, then so does the complement $\Omega \setminus A$;

3) if $A_1, A_2, \ldots$ is a sequence of sets in $\mathcal{F}$, then their union $A_1 \cup A_2 \cup \cdots$ also belongs to $\mathcal{F}$.

Example 1.1

Throughout this course $\mathbb{R}$ will denote the set of real numbers. The family of *Borel sets* $\mathcal{F} = \mathcal{B}(\mathbb{R})$ is a σ-field on $\mathbb{R}$. We recall that $\mathcal{B}(\mathbb{R})$ is the smallest σ-field containing all intervals in $\mathbb{R}$.

Definition 1.2

Let $\mathcal{F}$ be a σ-field on Ω. A *probability measure* P is a function

$$P : \mathcal{F} \to [0,1]$$

such that

1) $P(\Omega) = 1$;

2) if $A_1, A_2, \ldots$ are pairwise disjoint sets (that is, $A_i \cap A_j = \emptyset$ for $i \neq j$) belonging to $\mathcal{F}$, then

$$P(A_1 \cup A_2 \cup \cdots) = P(A_1) + P(A_2) + \cdots .$$

The triple $(\Omega, \mathcal{F}, P)$ is called a *probability space.* The sets belonging to $\mathcal{F}$ are called *events.* An event A is said to occur *almost surely* (a.s.) whenever $P(A) = 1$.

Example 1.2

We take the unit interval $\Omega = [0,1]$ with the σ-field $\mathcal{F} = \mathcal{B}([0,1])$ of Borel sets $B \subset [0,1]$, and *Lebesgue measure* $P =$ Leb on $[0,1]$. Then $(\Omega, \mathcal{F}, P)$ is a probability space. Recall that Leb is the unique measure defined on Borel sets such that

$$\text{Leb}[a,b] = b - a$$

for any interval $[a,b]$. (In fact Leb can be extended to a larger σ-field, but we shall need Borel sets only.)

Exercise 1.1

Show that if $A_1, A_2, \ldots$ is an *expanding* sequence of events, that is,

$$A_1 \subset A_2 \subset \cdots ,$$

then

$$P(A_1 \cup A_2 \cup \cdots) = \lim_{n\to\infty} P(A_n).$$

Similarly, if $A_1, A_2, \ldots$ is a *contracting* sequence of events, that is,

$$A_1 \supset A_2 \supset \cdots,$$

then

$$P(A_1 \cap A_2 \cap \cdots) = \lim_{n\to\infty} P(A_n).$$

Hint Write $A_1 \cup A_2 \cup \cdots$ as the union of a sequence of disjoint events: start with A_1, then add a disjoint set to obtain $A_1 \cup A_2$, then add a disjoint set again to obtain $A_1 \cup A_2 \cup A_3$, and so on. Now that you have a sequence of disjoint sets, you can use the definition of a probability measure. To deal with the product $A_1 \cap A_2 \cap \cdots$ write it as a union of some events with the aid of De Morgan's law.

Lemma 1.1 (Borel–Cantelli)

Let $A_1, A_2, \ldots$ be a sequence of events such that $P(A_1) + P(A_2) + \cdots < \infty$ and let $B_n = A_n \cup A_{n+1} \cup \cdots$. Then $P(B_1 \cap B_2 \cap \cdots) = 0$.

Exercise 1.2

Prove the Borel–Cantelli lemma above.

Hint $B_1, B_2, \ldots$ is a contracting sequence of events.

1.2 Random Variables

Definition 1.3

If $\mathcal{F}$ is a σ-field on Ω, then a function $\xi : \Omega \to \mathbb{R}$ is said to be $\mathcal{F}$*-measurable* if

$$\{\xi \in B\} \in \mathcal{F}$$

for every Borel set $B \in \mathcal{B}(\mathbb{R})$. If $(\Omega, \mathcal{F}, P)$ is a probability space, then such a function ξ is called a *random variable*.

Remark 1.1

A short-hand notation for events such as $\{\xi \in B\}$ will be used to avoid clutter. To be precise, we should write

$$\{\omega \in \Omega : \xi(\omega) \in B\}$$

in place of $\{\xi \in B\}$. Incidentally, $\{\xi \in B\}$ is just a convenient way of writing the inverse image $\xi^{-1}(B)$ of a set.

Definition 1.4

The σ-field $\sigma(\xi)$ *generated* by a random variable $\xi : \Omega \to \mathbb{R}$ consists of all sets of the form $\{\xi \in B\}$, where B is a Borel set in $\mathbb{R}$.

Definition 1.5

The σ-field $\sigma\{\xi_i : i \in I\}$ generated by a family $\{\xi_i : i \in I\}$ of random variables is defined to be the smallest σ-field containing all events of the form $\{\xi_i \in B\}$, where B is a Borel set in $\mathbb{R}$ and $i \in I$.

Exercise 1.3

We call $f : \mathbb{R} \to \mathbb{R}$ a *Borel function* if the inverse image $f^{-1}(B)$ of any Borel set B in $\mathbb{R}$ is a Borel set. Show that if f is a Borel function and ξ is a random variable, then the composition $f(\xi)$ is $\sigma(\xi)$-measurable.

Hint Consider the event $\{f(\xi) \in B\}$, where B is an arbitrary Borel set. Can this event be written as $\{\xi \in A\}$ for some Borel set A?

Lemma 1.2 (Doob–Dynkin)

Let ξ be a random variable. Then each $\sigma(\xi)$-measurable random variable η can be written as

$$\eta = f(\xi)$$

for some Borel function $f : \mathbb{R} \to \mathbb{R}$.

The proof of this highly non-trivial result will be omitted.

Definition 1.6

Every random variable $\xi : \Omega \to \mathbb{R}$ gives rise to a probability measure

$$P_\xi(B) = P\{\xi \in B\}$$

on $\mathbb{R}$ defined on the σ-field of Borel sets $B \in \mathcal{B}(\mathbb{R})$. We call P_ξ the *distribution* of ξ. The function $F_\xi : \mathbb{R} \to [0,1]$ defined by

$$F_\xi(x) = P\{\xi \leq x\}$$

is called the *distribution function* of ξ.

Exercise 1.4

Show that the distribution function F_ξ is non-decreasing, right-continuous, and

$$\lim_{x\to-\infty} F_\xi(x) = 0, \qquad \lim_{x\to+\infty} F_\xi(x) = 1.$$

Hint For example, to verify right-continuity show that $F_\xi(x_n) \to F_\xi(x)$ for any decreasing sequence x_n such that $x_n \to x$. You may find the results of Exercise 1.1 useful.

Definition 1.7

If there is a Borel function $f_\xi : \mathbb{R} \to \mathbb{R}$ such that for any Borel set $B \subset \mathbb{R}$

$$P\{\xi \in B\} = \int_B f_\xi(x)\, dx,$$

then ξ is said to be a random variable with *absolutely continuous distribution* and f_ξ is called the *density* of ξ. If there is a (finite or infinite) sequence of pairwise distinct real numbers $x_1, x_2, \ldots$ such that for any Borel set $B \subset \mathbb{R}$

$$P\{\xi \in B\} = \sum_{x_i \in B} P\{\xi = x_i\},$$

then ξ is said to have *discrete distribution* with values $x_1, x_2, \ldots$ and *mass* $P\{\xi = x_i\}$ at x_i.

Exercise 1.5

Suppose that ξ has continuous distribution with density f_ξ. Show that

$$\frac{d}{dx} F_\xi(x) = f_\xi(x)$$

if f_ξ is continuous at x.

Hint Express $F_\xi(x)$ as an integral of f_ξ.

Exercise 1.6

Show that if ξ has discrete distribution with values $x_1, x_2, \ldots$, then F_ξ is piecewise constant with jumps of size $P\{\xi = x_i\}$ at each x_i.

Hint The increment $F_\xi(t) - F_\xi(s)$ is equal to the total mass of the x_i's that belong to the interval $[s, t)$.

Definition 1.8

The *joint distribution* of several random variables $\xi_1, \ldots, \xi_n$ is a probability measure $P_{\xi_1,\ldots,\xi_n}$ on $\mathbb{R}^n$ such that

$$P_{\xi_1,\ldots,\xi_n}(B) = P\{(\xi_1, \ldots, \xi_n) \in B\}$$

for any Borel set B in $\mathbb{R}^n$. If there is a Borel function $f_{\xi_1,\ldots,\xi_n} : \mathbb{R}^n \to \mathbb{R}$ such that

$$P\{(\xi_1, \ldots, \xi_n) \in B\} = \int_B f_{\xi_1,\ldots,\xi_n}(x_1, \ldots, x_n)\, dx_1 \cdots dx_n$$

for any Borel set B in $\mathbb{R}^n$, then $f_{\xi_1,\ldots,\xi_n}$ is called the *joint density* of $\xi_1, \ldots, \xi_n$.

Definition 1.9

A random variable $\xi : \Omega \to \mathbb{R}$ is said to be *integrable* if

$$\int_\Omega |\xi|\, dP < \infty.$$

Then

$$E(\xi) = \int_\Omega \xi\, dP$$

exists and is called the *expectation* of ξ. The family of integrable random variables $\xi : \Omega \to \mathbb{R}$ will be denoted by L^1 or, in case of possible ambiguity, by $L^1(\Omega, \mathcal{F}, P)$.

Example 1.3

The *indicator function* 1_A of a set A is equal to 1 on A and 0 on the complement $\Omega \setminus A$ of A. For any event A

$$E(1_A) = \int_\Omega 1_A\, dP = P(A).$$

We say that $\eta : \Omega \to \mathbb{R}$ is a *step function* if

$$\eta = \sum_{i=1}^n \eta_i 1_{A_i},$$

where $\eta_1, \ldots, \eta_n$ are real numbers and $A_1, \ldots, A_n$ are pairwise disjoint events. Then

$$E(\eta) = \int_\Omega \eta\, dP = \sum_{i=1}^n \eta_i \int_\Omega 1_{A_i}\, dP = \sum_{i=1}^n \eta_i P(A_i).$$

Exercise 1.7

Show that for any Borel function $h : \mathbb{R} \to \mathbb{R}$ such that $h(\xi)$ is integrable

$$E(h(\xi)) = \int_{\mathbb{R}} h(x)\, dP_\xi(x).$$

Hint First verify the equality for step functions $h : \mathbb{R} \to \mathbb{R}$, then for non-negative ones by approximating them by step functions, and finally for arbitrary Borel functions by splitting them into positive and negative parts.

In particular, Exercise 1.7 implies that if ξ has an absolutely continuous distribution with density f_ξ, then

$$E(h(\xi)) = \int_{-\infty}^{+\infty} h(x)\, f_\xi(x)\, dx.$$

If ξ has a discrete distribution with (finitely or infinitely many) pairwise distinct values $x_1, x_2, \ldots$, then

$$E(h(\xi)) = \sum_i h(x_i)\, P\{\xi = x_i\}.$$

Definition 1.10

A random variable $\xi : \Omega \to \mathbb{R}$ is called *square integrable* if

$$\int_\Omega |\xi|^2\, dP < \infty.$$

Then the *variance* of ξ can be defined by

$$\operatorname{var}(\xi) = \int_\Omega (\xi - E(\xi))^2\, dP.$$

The family of square integrable random variables $\xi : \Omega \to \mathbb{R}$ will be denoted by $L^2(\Omega, \mathcal{F}, P)$ or, if no ambiguity is possible, simply by L^2.

Remark 1.2

The result in Exercise 1.8 below shows that we may write $E(\xi)$ in the definition of variance.

Exercise 1.8

Show that if ξ is a square integrable random variable, then it is integrable.

Hint Use the Schwarz inequality

$$[E(\xi\eta)]^2 \leq E(\xi^2)\, E(\eta^2) \tag{1.1}$$

with an appropriately chosen η.

Exercise 1.9

Show that if $\eta : \Omega \to [0, \infty)$ is a non-negative square integrable random variable, then

$$E(\eta^2) = 2\int_0^\infty tP(\eta > t)\, dt.$$

Hint Express $E(\eta^2)$ in terms of the distribution function $F_\eta(t)$ of η and then integrate by parts.

1.3 Conditional Probability and Independence

Definition 1.11

For any events $A, B \in \mathcal{F}$ such that $P(B) \neq 0$ the *conditional probability* of A given B is defined by

$$P(A|B) = \frac{P(A \cap B)}{P(B)}.$$

Exercise 1.10

Prove the *total probability formula*

$$P(A) = P(A|B_1)P(B_1) + P(A|B_2)P(B_2) + \cdots$$

for any event $A \in \mathcal{F}$ and any sequence of pairwise disjoint events $B_1, B_2, \ldots \in \mathcal{F}$ such that $B_1 \cup B_2 \cup \cdots = \Omega$ and $P(B_n) \neq 0$ for any n.

Hint $A = (A \cap B_1) \cup (A \cap B_2) \cup \cdots$.

Definition 1.12

Two events $A, B \in \mathcal{F}$ are called *independent* if

$$P(A \cap B) = P(A)P(B).$$

In general, we say that n events $A_1, \ldots, A_n \in \mathcal{F}$ are *independent* if

$$P(A_{i_1} \cap A_{i_2} \cap \cdots \cap A_{i_k}) = P(A_{i_i})P(A_{i_2}) \cdots P(A_{i_k})$$

for any indices $1 \leq i_1 < i_2 < \cdots < i_k \leq n$.

Exercise 1.11

Let $P(B) \neq 0$. Show that A and B are independent events if and only if $P(A|B) = P(A)$.

Hint If $P(B) \neq 0$, then you can divide by it.

Definition 1.13

Two random variables ξ and η are called *independent* if for any Borel sets $A, B \in \mathcal{B}(\mathbb{R})$ the two events

$$\{\xi \in A\} \quad \text{and} \quad \{\eta \in B\}$$

are independent. We say that n random variables $\xi_1, \ldots, \xi_n$ are *independent* if for any Borel sets $B_1, \ldots, B_n \in \mathcal{B}(\mathbb{R})$ the events

$$\{\xi_1 \in B_1\}, \ldots, \{\xi_n \in B_n\}$$

are independent. In general, a (finite or infinite) family of random variables is said to be *independent* if any finite number of random variables from this family are independent.

Proposition 1.1

If two integrable random variables $\xi, \eta : \Omega \to \mathbb{R}$ are independent, then they are *uncorrelated*, i.e.

$$E(\xi\eta) = E(\xi)E(\eta),$$

provided that the product $\xi\eta$ is also integrable. If $\xi_1, \ldots, \xi_n : \Omega \to \mathbb{R}$ are independent integrable random variables, then

$$E(\xi_1\xi_2 \cdots \xi_n) = E(\xi_1)E(\xi_2) \cdots E(\xi_n),$$

provided that the product $\xi_1\xi_2 \cdots \xi_n$ is also integrable.

Definition 1.14

Two σ-fields $\mathcal{G}$ and $\mathcal{H}$ contained in $\mathcal{F}$ are called *independent* if any two events

$$A \in \mathcal{G} \quad \text{and} \quad B \in \mathcal{H}$$

are independent. Similarly, any finite number of σ-fields $\mathcal{G}_1, \ldots, \mathcal{G}_n$ contained in $\mathcal{F}$ are *independent* if any n events

$$A_1 \in \mathcal{G}_1, \ldots, A_n \in \mathcal{G}_n$$

are independent. In general, a (finite or infinite) family of σ-fields is said to be *independent* if any finite number of them are independent.

Exercise 1.12

Show that two random variables ξ and η are independent if and only if the σ-fields $\sigma(\xi)$ and $\sigma(\eta)$ generated by them are independent.

Hint The events in $\sigma(\xi)$ and $\sigma(\eta)$ are of the form $\{\xi \in A\}$, and $\{\eta \in B\}$, where A and B are Borel sets.

Sometimes it is convenient to talk of independence for a combination of random variables and σ-fields.

Definition 1.15

We say that a random variable ξ is *independent* of a σ-field $\mathcal{G}$ if the σ-fields

$$\sigma(\xi) \quad \text{and} \quad \mathcal{G}$$

are independent. This can be extended to any (finite or infinite) family consisting of random variables or σ-fields or a combination of them both. Namely, such a family is called *independent* if for any finite number of random variables $\xi_1, \ldots, \xi_m$ and σ-fields $\mathcal{G}_1, \ldots, \mathcal{G}_n$ from this family the σ-fields

$$\sigma(\xi_1), \ldots, \sigma(\xi_m), \mathcal{G}_1, \ldots, \mathcal{G}_n$$

are independent.

1.4 Solutions

Solution 1.1

If $A_1 \subset A_2 \subset \cdots$, then

$$A_1 \cup A_2 \cup \cdots = A_1 \cup (A_2 \setminus A_1) \cup (A_3 \setminus A_2) \cup \cdots ,$$

where the sets $A_1, A_2 \setminus A_1, A_3 \setminus A_2, \ldots$ are pairwise disjoint. Therefore, by the definition of probability measure

$$\begin{aligned} P(A_1 \cup A_2 \cup \cdots) &= P(A_1 \cup (A_2 \setminus A_1) \cup (A_3 \setminus A_2) \cup \cdots) \\ &= P(A_1) + P(A_2 \setminus A_1) + P(A_3 \setminus A_2) + \cdots \\ &= \lim_{n\to\infty} P(A_n). \end{aligned}$$

The last equality holds because the partial sums in the series above are

$$\begin{aligned} P(A_1) + P(A_2 \setminus A_1) + \cdots + P(A_n \setminus A_{n-1}) &= P(A_1 \cup \cdots \cup A_n) \\ &= P(A_n). \end{aligned}$$

If $A_1 \supset A_2 \supset \cdots$, then the equality

$$P(A_1 \cap A_2 \cap \cdots) = \lim_{n\to\infty} P(A_n)$$

follows by taking the complements of A_n and applying De Morgan's law

$$\Omega \setminus (A_1 \cap A_2 \cap \cdots) = (\Omega \setminus A_1) \cup (\Omega \setminus A_2) \cup \cdots .$$

Solution 1.2

Since B_n is a contracting sequence of events, the results of Exercise 1.1 imply that

$$\begin{aligned} P(B_1 \cap B_2 \cap \cdots) &= \lim_{n\to\infty} P(B_n) \\ &= \lim_{n\to\infty} P(A_n \cup A_{n+1} \cup \cdots) \\ &\leq \lim_{n\to\infty} (P(A_n) + P(A_{n+1}) + \cdots) \\ &= 0. \end{aligned}$$

The last equality holds because the series $\sum_{n=1}^{\infty} P(A_n)$ is convergent. The inequality above holds by the subadditivity property

$$P(A_n \cup A_{n+1} \cup \cdots) \leq P(A_n) + P(A_{n+1}) + \cdots .$$

It follows that $P(B_1 \cap B_2 \cap \cdots) = 0$.

Solution 1.3

If B is a Borel set in $\mathbb{R}$ and $f : \mathbb{R} \to \mathbb{R}$ is a Borel function, then $f^{-1}(B)$ is also a Borel set. Therefore

$$\{f(\xi) \in B\} = \{\xi \in f^{-1}(B)\}$$

belongs to the σ-field $\sigma(\xi)$ generated by ξ. It follows that the composition $f(\xi)$ is $\sigma(\xi)$-measurable.

Solution 1.4

If $x \leq y$, then $\{\xi \leq x\} \subset \{\xi \leq y\}$, so

$$F_\xi(x) = P\{\xi \leq x\} \leq P\{\xi \leq y\} = F_\xi(y).$$

This means that F_ξ is non-decreasing.

Next, we take any sequence $x_1 \geq x_2 \geq \cdots$ and put

$$\lim_{n\to\infty} x_n = x.$$

Then the events

$$\{\xi \leq x_1\} \supset \{\xi \leq x_2\} \supset \cdots$$

form a contracting sequence with intersection

$$\{\xi \leq x\} = \{\xi \leq x_1\} \cap \{\xi \leq x_2\} \cap \cdots.$$

It follows by Exercise 1.1 that

$$F_\xi(x) = P\{\xi \leq x\} = \lim_{n\to\infty} P\{\xi \leq x_n\} = \lim_{n\to\infty} F_\xi(x_n).$$

This proves that F_ξ is right-continuous.

Since the events

$$\{\xi \leq -1\} \supset \{\xi \leq -2\} \supset \cdots$$

form a contracting sequence with intersection $\emptyset$ and

$$\{\xi \leq 1\} \subset \{\xi \leq 2\} \subset \cdots$$

form an expanding sequence with union Ω, it follows by Exercise 1.1 that

$$\lim_{x\to-\infty} F_\xi(x) = \lim_{n\to\infty} F_\xi(-n) = \lim_{n\to\infty} P\{\xi \leq -n\} = P(\emptyset) = 0,$$

$$\lim_{x\to\infty} F_\xi(x) = \lim_{n\to\infty} F_\xi(n) = \lim_{n\to\infty} P\{\xi \leq n\} = P(\Omega) = 1,$$

since F_ξ is non-decreasing.

Solution 1.5

If ξ has a density f_ξ, then the distribution function F_ξ can be written as

$$F_\xi(x) = P\{\xi \leq x\} = \int_{-\infty}^{x} f_\xi(y)\, dy.$$

Therefore, if f_ξ is continuous at x, then F_ξ is differentiable at x and

$$\frac{d}{dx} F_\xi(x) = f_\xi(x).$$

Solution 1.6

Suppose that ξ has discrete distribution with values $x_1, x_2, \ldots$. If $s < t$ are real numbers such that $x_i \notin (s,t]$ for any i, then

$$F_\xi(t) - F_\xi(s) = P\{\xi \le t\} - P\{\xi \le s\} = P\{\xi \in (s,t]\} = 0,$$

i.e. $F_\xi(s) = F_\xi(t)$. But if there is exactly one i such that $x_i \in (s,t]$, then

$$F_\xi(t) - F_\xi(s) = P\{\xi \le t\} - P\{\xi \le s\} = P\{\xi \in (s,t]\} = P\{\xi = x_i\}.$$

This means that F_ξ is constant on any interval not containing any of the numbers $x_1, x_2, \ldots$ and has a jump of size $P\{\xi = x_i\}$ at each x_i.

Solution 1.7

If h is a step function,

$$h = \sum_{i=1}^{n} h_i 1_{A_i},$$

where $h_1, \ldots, h_n$ are real numbers and $A_1, \ldots, A_n$ are pairwise disjoint Borel sets covering $\mathbb{R}$, then

$$\begin{aligned} E(h(\xi)) &= \sum_{i=1}^{n} h_i E(1_{A_i}(\xi)) = \sum_{i=1}^{n} h_i P\{\xi \in A_i\} \\ &= \sum_{i=1}^{n} h_i P_\xi(A_i) = \sum_{i=1}^{n} \int_{A_i} h(x)\, dP_\xi(x) = \int_{\mathbb{R}} h(x)\, dP_\xi(x). \end{aligned}$$

Next, any non-negative Borel function h can be approximated by a non-decreasing sequence of step functions. For such an h the result follows by the monotone convergence of integrals. Finally, this implies the desired equality for all Borel functions h, since each can be split into its positive and negative parts, $h = h^+ - h^-$, where $h^+, h^- \ge 0$.

Solution 1.8

By the Schwarz inequality (1.1) with $\eta = 1$, if ξ is square integrable, then

$$[E(|\xi|)]^2 = [E(1\,|\xi|)]^2 \le E(1^2)\, E(\xi^2) = E(\xi^2) < \infty,$$

i.e. ξ is integrable.

Solution 1.9

Let $F(t) = P\{\eta \le t\}$ be the distribution function of η. Then

$$E(\eta^2) = \int_0^\infty t^2 dF(t).$$

Since $P(\eta > t) = 1 - F(t)$, we need to show that

$$\int_0^\infty t^2\, dF(t) = 2\int_0^\infty t\,(1 - F(t))\; dt \tag{1.2}$$

First, let us establish a version of (1.2) with ∞ replaced by a finite number a. Integrating by parts, we obtain

$$\begin{aligned} \int_0^a t^2 dF(t) &= \int_0^a t^2 d(F(t) - 1) \\ &= t^2(F(t) - 1)\big|_0^a - 2\int_0^a t(F(t) - 1)\, dt \\ &= -a^2(1 - F(a)) + 2\int_0^a t(1 - F(t))dt. \end{aligned} \tag{1.3}$$

We see that (1.2) follows from (1.3), provided that

$$a^2\,(1 - F(a)) \to 0, \quad \text{as } a \to \infty. \tag{1.4}$$

But

$$0 \leq a^2\,(1 - F(a)) = a^2 P(\eta > a) \leq (n+1)^2 P(\eta > n) \leq 4n^2 P(\eta \geq n),$$

where n is the integer part of a, and

$$E(\eta^2) = \sum_{k=0}^{\infty} \int_{\{k \leq \eta \leq k+1\}} \eta^2 dP < \infty.$$

Hence,

$$n^2 P\,(\eta \geq n) \leq \int_{\{\eta \geq n\}} \eta^2\, dP = \sum_{k=n}^{\infty} \int_{\{k \leq \eta < k+1\}} \eta^2\, dP \to 0 \tag{1.5}$$

as $n \to \infty$, which proves (1.4).

Solution 1.10

Since $B_1 \cup B_2 \cup \cdots = \Omega$,

$$A = A \cap (B_1 \cup B_2 \cup \cdots) = (A \cap B_1) \cup (A \cap B_2) \cup \cdots,$$

where

$$(A \cap B_i) \cap (A \cap B_j) = A \cap (B_i \cap B_j) = A \cap \emptyset = \emptyset.$$

By countable additivity

$$\begin{aligned} P(A) &= P\left((A \cap B_1) \cup (A \cap B_2) \cup \cdots\right) \\ &= P\,(A \cap B_1) + P\,(A \cap B_2) + \cdots \\ &= P\,(A|B_1)\, P(B_1) + P\,(A|B_2)\, P(B_2) + \cdots. \end{aligned}$$

Solution 1.11

If $P(B) \neq 0$, then A and B are independent if and only if

$$P(A) = \frac{P(A \cap B)}{P(B)}.$$

In turn, this equality holds if and only if $P(A) = P(A|B)$.

Solution 1.12

The σ-fields $\sigma(\xi)$ and $\sigma(\eta)$ consist, respectively, of events of the form

$$\{\xi \in A\} \quad \text{and} \quad \{\eta \in B\},$$

where A and B are Borel sets in $\mathbb{R}$. Therefore, $\sigma(\xi)$ and $\sigma(\eta)$ are independent if and only if the events $\{\xi \in A\}$, and $\{\eta \in B\}$ are independent for any Borel sets A and B, which in turn is equivalent to ξ and η being independent.

2
Conditional Expectation

Conditional expectation is a crucial tool in the study of stochastic processes. It is therefore important to develop the necessary intuition behind this notion, the definition of which may appear somewhat abstract at first. This chapter is designed to help the beginner by leading him or her step by step through several special cases, which become increasingly involved, culminating at the general definition of conditional expectation. Many varied examples and exercises are provided to aid the reader's understanding.

2.1 Conditioning on an Event

The first and simplest case to consider is that of the conditional expectation $E(\xi|B)$ of a random variable ξ given an event B.

Definition 2.1

For any integrable random variable ξ and any event $B \in \mathcal{F}$ such that $P(B) \neq 0$ the *conditional expectation* of ξ given B is defined by

$$E(\xi|B) = \frac{1}{P(B)} \int_B \xi \, dP.$$

Example 2.1

Three coins, 10p, 20p and 50p are tossed. The values of those coins that land heads up are added to work out the total amount ξ. What is the expected total amount ξ given that two coins have landed heads up?

Let B denote the event that two coins have landed heads up. We want to find $E(\xi|B)$. Clearly, B consists of three elements,

$$B = \{\mathrm{HHT}, \mathrm{HTH}, \mathrm{THH}\},$$

each having the same probability $\frac{1}{8}$. (Here H stands for heads and T for tails.) The corresponding values of ξ are

$$\begin{aligned}\xi(\mathrm{HHT}) &= 10 + 20 = 30, \\ \xi(\mathrm{HTH}) &= 10 + 50 = 60, \\ \xi(\mathrm{THH}) &= 20 + 50 = 70.\end{aligned}$$

Therefore

$$E(\xi|B) = \frac{1}{P(B)} \int_B \xi \, dP = \frac{1}{\frac{3}{8}} \left(\frac{30}{8} + \frac{60}{8} + \frac{70}{8} \right) = 53\tfrac{1}{3}.$$

Exercise 2.1

Show that $E(\xi|\Omega) = E(\xi)$.

Hint The definition of $E(\xi)$ involves an integral and so does the definition of $E(\xi|\Omega)$. How are these integrals related?

Exercise 2.2

Show that if

$$1_A(\omega) = \begin{cases} 1 & \text{for } \omega \in A \\ 0 & \text{for } \omega \notin A \end{cases}$$

(the *indicator function* of A), then

$$E(1_A|B) = P(A|B),$$

where

$$P(A|B) = \frac{P(A \cap B)}{P(B)}$$

is the *conditional probability* of A given B.

Hint Write $\int_B 1_A \, dP$ as $P(A \cap B)$.

2.2 Conditioning on a Discrete Random Variable

The next step towards the general definition of conditional expectation involves conditioning by a discrete random variable η with possible values $y_1, y_2, \ldots$ such that $P\{\eta = y_n\} \neq 0$ for each n. Finding out the value of η amounts to finding out which of the events $\{\eta = y_n\}$ has occurred or not. Conditioning by η should therefore be the same as conditioning by the events $\{\eta = y_n\}$. Because we do not know in advance which of these events will occur, we need to consider all possibilities, involving a sequence of conditional expectations

$$E(\xi|\{\eta = y_1\}), E(\xi|\{\eta = y_2\}), \ldots .$$

A convenient way of doing this is to construct a new discrete random variable constant and equal to $E(\xi|\{\eta = y_n\})$ on each of the sets $\{\eta = y_n\}$. This leads us to the next definition.

Definition 2.2

Let ξ be an integrable random variable and let η be a discrete random variable as above. Then the *conditional expectation* of ξ given η is defined to be a random variable $E(\xi|\eta)$ such that

$$E(\xi|\eta)(\omega) = E(\xi|\{\eta = y_n\}) \quad \text{if } \eta(\omega) = y_n$$

for any $n = 1, 2, \ldots$.

Example 2.2

Three coins, 10p, 20p and 50p are tossed as in Example 2.1. What is the conditional expectation $E(\xi|\eta)$ of the total amount ξ shown by the three coins given the total amount η shown by the 10p and 20p coins only?

Clearly, η is a discrete random variable with four possible values: 0, 10, 20 and 30. We find the four corresponding conditional expectations in a similar way as in Example 2.1:

$$E(\xi|\{\eta = 0\}) = 25, \qquad E(\xi|\{\eta = 10\}) = 35,$$
$$E(\xi|\{\eta = 20\}) = 45, \qquad E(\xi|\{\eta = 30\}) = 55.$$

Therefore

$$E(\xi|\eta)(\omega) = \begin{cases} 25 & \text{if } \eta(\omega) = 0, \\ 35 & \text{if } \eta(\omega) = 10, \\ 45 & \text{if } \eta(\omega) = 20, \\ 55 & \text{if } \eta(\omega) = 30. \end{cases}$$

Example 2.3

Take $\Omega = [0,1]$ with the σ-field of Borel sets and P the Lebesgue measure on $[0,1]$. We shall find $E(\xi|\eta)$ for

$$\xi(x) = 2x^2, \qquad \eta(x) = \begin{cases} 1 & \text{if } x \in [0, \frac{1}{3}], \\ 2 & \text{if } x \in (\frac{1}{3}, \frac{2}{3}), \\ 0 & \text{if } x \in (\frac{2}{3}, 1]. \end{cases}$$

Clearly, η is discrete with three possible values $1, 2, 0$. The corresponding events are

$$\begin{aligned} \{\eta = 1\} &= [0, \tfrac{1}{3}], \\ \{\eta = 2\} &= (\tfrac{1}{3}, \tfrac{2}{3}), \\ \{\eta = 0\} &= (\tfrac{2}{3}, 1]. \end{aligned}$$

For $x \in [0, \frac{1}{3}]$

$$E(\xi|\eta)(x) = E(\xi|[0, \tfrac{1}{3}]) = \frac{1}{\frac{1}{3}} \int_0^{\frac{1}{3}} 2x^2 dx = \frac{2}{27}.$$

For $x \in (\frac{1}{3}, \frac{2}{3})$

$$E(\xi|\eta)(x) = E(\xi|(\tfrac{1}{3}, \tfrac{2}{3})) = \frac{1}{\frac{1}{3}} \int_{\frac{1}{3}}^{\frac{2}{3}} 2x^2 dx = \frac{14}{27}.$$

And for $x \in (\frac{2}{3}, 1]$

$$E(\xi|\eta)(x) = E(\xi|(\tfrac{2}{3}, 1]) = \frac{1}{\frac{1}{3}} \int_{\frac{2}{3}}^{1} 2x^2 dx = \frac{38}{27}.$$

The graph of $E(\xi|\eta)$ is shown in Figure 2.1 together with those of ξ and η.

Exercise 2.3

Show that if η is a constant function, then $E(\xi|\eta)$ is constant and equal to $E(\xi)$.

Hint The event $\{\eta = c\}$ must be $\emptyset$ or Ω for any $c \in \mathbb{R}$.

Exercise 2.4

Show that

$$E(1_A|1_B)(\omega) = \begin{cases} P(A|B) & \text{if } \omega \in B \\ P(A|\Omega \setminus B) & \text{if } \omega \notin B \end{cases}$$

for any B such that $1 \neq P(B) \neq 0$.

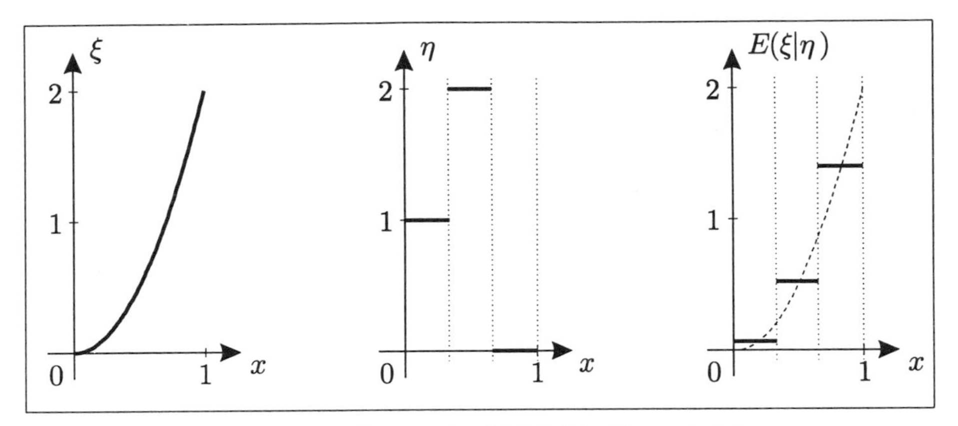

Figure 2.1. The graph of $E(\xi|\eta)$ in Example 2.3

Hint How many different values does 1_B take? What are the sets on which these values are taken?

Exercise 2.5

Assuming that η is a discrete random variable, show that

$$E(E(\xi|\eta)) = E(\xi).$$

Hint Observe that

$$\int_B E(\xi|\eta)\, dP = \int_B E(\xi)\, dP$$

for any event B on which η is constant. The desired equality can be obtained by covering Ω by countably many disjoint events of this kind.

Proposition 2.1

If ξ is an integrable random variable and η is a discrete random variable, then

1) $E(\xi|\eta)$ is $\sigma(\eta)$-measurable;

2) For any $A \in \sigma(\eta)$

$$\int_A E(\xi|\eta)\, dP = \int_A \xi\, dP. \tag{2.1}$$

Proof

Suppose that η has pairwise distinct values $y_1, y_2, \ldots$. Then the events

$$\{\eta = y_1\}, \{\eta = y_2\}, \ldots$$

are pairwise disjoint and cover Ω. The σ-field $\sigma(\eta)$ is generated by these events, in fact every $A \in \sigma(\eta)$ is a countable union of sets of the form $\{\eta = y_n\}$. Because $E(\xi|\eta)$ is constant on each of these sets, it must be $\sigma(\eta)$-measurable.

For each n we have

$$\begin{aligned}\int_{\{\eta=y_n\}} E(\xi|\eta)\, dP &= \int_{\{\eta=y_n\}} E(\xi|\{\eta = y_n\})\, dP \\ &= \int_{\{\eta=y_n\}} \xi\, dP.\end{aligned}$$

Since each $A \in \sigma(\eta)$ is a countable union of sets of the form $\{\eta = y_n\}$, which are pairwise disjoint, it follows that

$$\int_A E(\xi|\eta)\, dP = \int_A \xi\, dP,$$

as required. □

2.3 Conditioning on an Arbitrary Random Variable

Properties 1) and 2) in Proposition 2.1 provide the key to the definition of the conditional expectation given an arbitrary random variable η.

Definition 2.3

Let ξ be an integrable random variable and let η be an arbitrary random variable. Then the *conditional expectation* of ξ given η is defined to be a random variable $E(\xi|\eta)$ such that

1) $E(\xi|\eta)$ is $\sigma(\eta)$-measurable;
2) For any $A \in \sigma(\eta)$

$$\int_A E(\xi|\eta)\, dP = \int_A \xi\, dP.$$

Remark 2.1

We can also define the *conditional probability* of an event $A \in \mathcal{F}$ given η by

$$P(A|\eta) = E(1_A|\eta),$$

where 1_A is the indicator function of A.

Do the conditions of Definition 2.3 characterize $E(\xi|\eta)$ uniquely? The lemma below implies that $E(\xi|\eta)$ is defined to within equality on a set of full measure. Namely,

$$\text{if } \xi = \xi' \text{ a.s., then } E(\xi|\eta) = E(\xi'|\eta) \text{ a.s.} \tag{2.2}$$

The existence of $E(\xi|\eta)$ will be discussed later in this chapter.

Lemma 2.1

Let $(\Omega, \mathcal{F}, P)$ be a probability space and let $\mathcal{G}$ be a σ-field contained in $\mathcal{F}$. If ξ is a $\mathcal{G}$-measurable random variable and for any $B \in \mathcal{G}$

$$\int_B \xi \, dP = 0,$$

then $\xi = 0$ a.s.

Proof

Observe that $P\{\xi \geq \varepsilon\} = 0$ for any $\varepsilon > 0$ because

$$0 \leq \varepsilon P\{\xi \geq \varepsilon\} = \int_{\{\xi \geq \varepsilon\}} \varepsilon \, dP \leq \int_{\{\xi \geq \varepsilon\}} \xi \, dP = 0.$$

The last equality holds, since $\{\xi \geq \varepsilon\} \in \mathcal{G}$. Similarly, $P\{\xi \leq -\varepsilon\} = 0$ for any $\varepsilon > 0$. As a consequence,

$$P\{-\varepsilon < \xi < \varepsilon\} = 1$$

for any $\varepsilon > 0$.

Let us put

$$A_n = \left\{-\tfrac{1}{n} < \xi < \tfrac{1}{n}\right\}.$$

Then $P(A_n) = 1$ and

$$\{\xi = 0\} = \bigcap_{n=1}^{\infty} A_n.$$

Because the A_n form a contracting sequence of events, it follows that

$$P\{\xi = 0\} = \lim_{n \to \infty} P(A_n) = 1,$$

as required. □

One difficulty involved in Definition 2.3 is that no explicit formula for $E(\xi|\eta)$ is given. If such a formula is known, then it is usually fairly easy to verify conditions 1) and 2). But how do you find it in the first place? The examples and exercises below are designed to show how to tackle this problem in concrete cases.

Example 2.4

Take $\Omega = [0,1]$ with the σ-field of Borel sets and P the Lebesgue measure on $[0,1]$. We shall find $E(\xi|\eta)$ for

$$\xi(x) = 2x^2, \qquad \eta(x) = \begin{cases} 2 & \text{if } x \in [0,\frac{1}{2}), \\ x & \text{if } x \in [\frac{1}{2},1]. \end{cases}$$

Here η is no longer discrete and the general Definition 2.3 should be used.

First we need to describe the σ-field $\sigma(\eta)$. For any Borel set $B \subset [\frac{1}{2},1]$ we have

$$B = \{\eta \in B\} \in \sigma(\eta)$$

and

$$[0,\tfrac{1}{2}) \cup B = \{\eta \in B\} \cup \{\eta = 2\} \in \sigma(\eta).$$

In fact sets of these two kinds exhaust all elements of $\sigma(\eta)$. The inverse image $\{\eta \in C\}$ of any Borel set $C \subset \mathbb{R}$ is of the first kind if $2 \notin C$ and of the second kind if $2 \in C$.

If $E(\xi|\eta)$ is to be $\sigma(\eta)$-measurable, it must be constant on $[0,\frac{1}{2})$ because η is. If for any $x \in [0,\frac{1}{2})$

$$\begin{aligned} E(\xi|\eta)(x) &= E(\xi|[0,\tfrac{1}{2})) \\ &= \frac{1}{P([0,\frac{1}{2}))}\int_{[0,\frac{1}{2})} \xi(x)\,dx \\ &= \frac{1}{\frac{1}{2}}\int_0^{\frac{1}{2}} 2x^2\,dx \\ &= \frac{1}{6}, \end{aligned}$$

then

$$\int_{[0,\frac{1}{2})} E(\xi|\eta)(x)\,dx = \int_{[0,\frac{1}{2})} \xi(x)\,dx,$$

i.e. condition 2) of Definition 2.3 will be satisfied for $A = [0,\frac{1}{2})$.

Moreover, if $E(\xi|\eta) = \xi$ on $[\frac{1}{2},1]$, then of course

$$\int_B E(\xi|\eta)(x)\,dx = \int_B \xi(x)\,dx$$

for any Borel set $B \subset [\frac{1}{2}, 1]$.

Therefore, we have found that

$$E(\xi|\eta)(x) = \begin{cases} \frac{1}{6} & \text{if } x \in [0, \frac{1}{2}), \\ 2x^2 & \text{if } x \in [\frac{1}{2}, 1]. \end{cases}$$

Because every element of $\sigma(\eta)$ is of the form B or $[0, \frac{1}{2}) \cup B$, where $B \subset [\frac{1}{2}, 1]$ is a Borel set, it follows immediately that conditions 1) and 2) of Definition 2.3 are satisfied. The graph of $E(\xi|\eta)$ is presented in Figure 2.2 along with those of ξ and η.

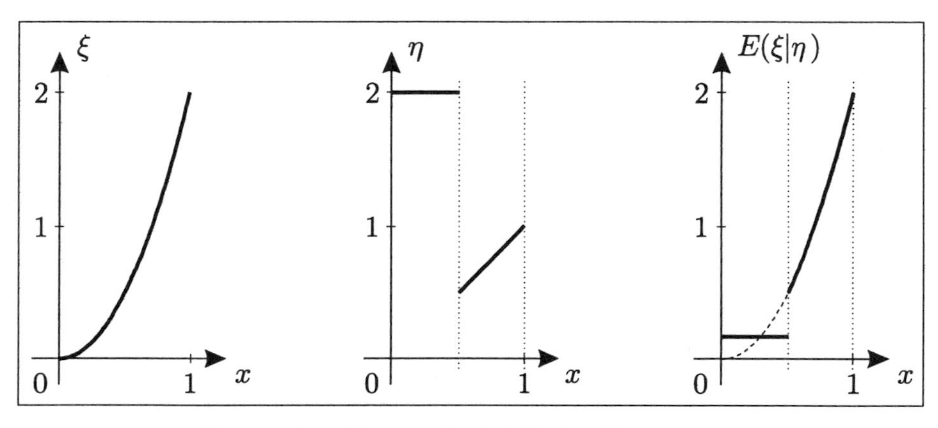

Figure 2.2. The graph of $E(\xi|\eta)$ in Example 2.4

Exercise 2.6

Let $\Omega = [0, 1]$ with Lebesgue measure as in Example 2.4. Find the conditional expectation $E(\xi|\eta)$ if

$$\xi(x) = 2x^2, \qquad \eta = 1 - |2x - 1|.$$

Hint First describe the σ-field generated by η. Observe that η is symmetric about $\frac{1}{2}$. What does it tell you about the sets in $\sigma(\eta)$? What does it tell you about $E(\xi|\eta)$ if it is to be $\sigma(\eta)$-measurable? Does it need to be symmetric as well? For any A in $\sigma(\eta)$ try to transform $\int_A \xi\, dP$ to make the integrand symmetric.

Exercise 2.7

Let Ω be the unit square $[0, 1] \times [0, 1]$ with the σ-field of Borel sets and P the Lebesgue measure on $[0, 1] \times [0, 1]$. Suppose that ξ and η are random variables on Ω with joint density

$$f_{\xi,\eta}(x, y) = x + y$$

for any $x, y \in [0,1]$, and $f_{\xi,\eta}(x,y) = 0$ otherwise. Show that

$$E(\xi|\eta) = \frac{2+3\eta}{3+6\eta}.$$

Hint It suffices (why?) to show that for any Borel set B

$$\int_{\{\eta\in B\}} \xi \, dP = \int_{\{\eta\in B\}} \frac{2+3\eta}{3+6\eta} \, dP.$$

Try to express each side of this equality as an integral over the square $[0,1] \times [0,1]$ using the joint density $f_{\xi,\eta}(x,y)$.

Exercise 2.8

Let Ω be the unit square $[0,1]\times[0,1]$ with Lebesgue measure as in Exercise 2.7. Find $E(\xi|\eta)$ if ξ and η are random variables on Ω with joint density

$$f_{\xi,\eta}(x,y) = \frac{3}{2}\left(x^2 + y^2\right)$$

for any $x, y \in [0,1]$, and $f_{\xi,\eta}(x,y) = 0$ otherwise.

Hint This is slightly harder than Exercise 2.7 because here we have to derive a formula for the conditional expectation. Study the solution to Exercise 2.7 to find a way of obtaining such a formula.

Exercise 2.9

Let Ω be the unit disc $\left\{(x,y) : x^2 + y^2 \leq 1\right\}$ with the σ-field of Borel sets and P the Lebesgue measure on the disc normalized so that $P(\Omega) = 1$, i.e.

$$P(A) = \frac{1}{\pi} \iint_A dx \, dy$$

for any Borel set $A \subset \Omega$. Suppose that ξ and η are the projections onto the x and y axes,

$$\xi(x,y) = x, \qquad \eta(x,y) = y$$

for any $(x,y) \in \Omega$. Find $E\left(\xi^2|\eta\right)$.

Hint What is the joint density of ξ and η? Use this density to transform the integral

$$\int_{\{\eta\in B\}} \xi^2 \, dP$$

for an arbitrary Borel set B so that the integrand becomes a function of η. How is this function of η related to $E\left(\xi^2|\eta\right)$?

2.4 Conditioning on a σ-Field

We are now in a position to make the final step towards the general definition of conditional expectation. It is based on the observation that $E(\xi|\eta)$ depends only on the σ-field $\sigma(\eta)$ generated by η, rather than on the actual values of η.

Proposition 2.2

If $\sigma(\eta) = \sigma(\eta')$, then $E(\xi|\eta) = E(\xi|\eta')$ a.s. (Compare this with (2.2).)

Proof

This is an immediate consequence of Lemma 2.1. □

Because of Proposition 2.2 it is reasonable to talk of conditional expectation given a σ-field. The definition below differs from Definition 2.3 only by using an arbitrary σ-field $\mathcal{G}$ in place of a σ-field $\sigma(\eta)$ generated by a random variable η.

Definition 2.4

Let ξ be an integrable random variable on a probability space $(\Omega, \mathcal{F}, P)$, and let $\mathcal{G}$ be a σ-field contained in $\mathcal{F}$. Then the *conditional expectation* of ξ given $\mathcal{G}$ is defined to be a random variable $E(\xi|\mathcal{G})$ such that

1) $E(\xi|\mathcal{G})$ is $\mathcal{G}$-measurable;
2) For any $A \in \mathcal{G}$
$$\int_A E(\xi|\mathcal{G})\, dP = \int_A \xi\, dP. \tag{2.3}$$

Remark 2.2

The *conditional probability* of an event $A \in \mathcal{F}$ given a σ-field $\mathcal{G}$ can be defined by
$$P(A|\mathcal{G}) = E(1_A|\mathcal{G}),$$
where 1_A is the indicator function of A.

The notion of conditional expectation with respect to a σ-field extends conditioning on a random variable η in the sense that
$$E(\xi|\sigma(\eta)) = E(\xi|\eta),$$
where $\sigma(\eta)$ is the σ-field generated by η.

Proposition 2.3

$E(\xi|\mathcal{G})$ exists and is unique in the sense that if $\xi = \xi'$ a.s., then $E(\xi|\mathcal{G}) = E(\xi'|\mathcal{G})$ a.s.

Proof

Existence and uniqueness follow, respectively, from Theorem 2.1 below and Lemma 2.1. □

Theorem 2.1 (Radon–Nikodym)

Let $(\Omega, \mathcal{F}, P)$ be a probability space and let $\mathcal{G}$ be a σ-field contained in $\mathcal{F}$. Then for any random variable ξ there exists a $\mathcal{G}$-measurable random variable ζ such that

$$\int_A \xi \, dP = \int_A \zeta \, dP$$

for each $A \in \mathcal{G}$.

The Radon–Nikodym theorem is important from a theoretical point of view. However, in practice there are usually other ways of establishing the existence of conditional expectation, for example, by finding an explicit formula, as in the examples and exercises in the previous section. The proof of the Radon–Nikodym theorem is beyond the scope of this course and is omitted.

Exercise 2.10

Show that if $\mathcal{G} = \{\emptyset, \Omega\}$, then $E(\xi|\mathcal{G}) = E(\xi)$ a.s.

Hint What random variables are $\mathcal{G}$-measurable if $\mathcal{G} = \{\emptyset, \Omega\}$?

Exercise 2.11

Show that if ξ is $\mathcal{G}$-measurable, then $E(\xi|\mathcal{G}) = \xi$ a.s.

Hint The conditions of Definition 2.4 are trivially satisfied by ξ if ξ is $\mathcal{G}$-measurable.

Exercise 2.12

Show that if $B \in \mathcal{G}$, then

$$E\left(E\left(\xi|\mathcal{G}\right)|B\right) = E\left(\xi|B\right).$$

Hint The conditional expectation on either side of the equality involves an integral over B. How are these integrals related to one another?

2.5 General Properties

Proposition 2.4

Conditional expectation has the following properties:

1) $E(a\xi + b\zeta|\mathcal{G}) = aE(\xi|\mathcal{G}) + bE(\zeta|\mathcal{G})$ (*linearity*);
2) $E(E(\xi|\mathcal{G})) = E(\xi)$;
3) $E(\xi\zeta|\mathcal{G}) = \xi E(\zeta|\mathcal{G})$ if ξ is $\mathcal{G}$-measurable (*taking out what is known*);
4) $E(\xi|\mathcal{G}) = E(\xi)$ if ξ is independent of $\mathcal{G}$ (*an independent condition drops out*);
5) $E(E(\xi|\mathcal{G})|\mathcal{H}) = E(\xi|\mathcal{H})$ if $\mathcal{H} \subset \mathcal{G}$ (*tower property*);
6) If $\xi \geq 0$, then $E(\xi|\mathcal{G}) \geq 0$ (*positivity*).

Here a, b are arbitrary real numbers, ξ, ζ are integrable random variables on a probability space $(\Omega, \mathcal{F}, P)$ and $\mathcal{G}, \mathcal{H}$ are σ-fields on Ω contained in $\mathcal{F}$. In 3) we also assume that the product $\xi\zeta$ is integrable. All equalities and the inequalities in 6) hold P-a.s.

Proof

1) For any $B \in \mathcal{G}$

$$\begin{aligned}\int_B (aE(\xi|\mathcal{G}) + bE(\zeta|\mathcal{G}))\, dP &= a\int_B E(\xi|\mathcal{G})\, dP + b\int_B E(\zeta|\mathcal{G})\, dP \\ &= a\int_B \xi\, dP + b\int_B \zeta\, dP \\ &= \int_B (a\xi + b\zeta)\, dP.\end{aligned}$$

By uniqueness this proves the desired equality.

2) This follows by putting $B = \Omega$ in (2.3). Also, 2) is a special case of 5) when $\mathcal{H} = \{\emptyset, \Omega\}$.

3) We first verify the result for $\xi = 1_A$, where $A \in \mathcal{G}$. In this case

$$\begin{aligned}\int_B 1_A E(\eta|\mathcal{G})\, dP &= \int_{A\cap B} E(\eta|\mathcal{G})\, dP \\ &= \int_{A\cap B} \eta\, dP \\ &= \int_B 1_A \eta\, dP\end{aligned}$$

for any $B \in \mathcal{G}$, which implies that

$$1_A E(\eta|\mathcal{G}) = E(1_A \eta|\mathcal{G})$$

by uniqueness. In a similar way we obtain the result if ξ is a $\mathcal{G}$-measurable step function,

$$\xi = \sum_{j=1}^{m} a_j 1_{A_j},$$

where $A_j \in \mathcal{G}$ for $j = 1, \ldots, m$. Finally, the result in the general case follows by approximating ξ by $\mathcal{G}$-measurable step functions.

4) Since ξ is independent of $\mathcal{G}$, the random variables ξ and 1_B are independent for any $B \in \mathcal{G}$. It follows by Proposition 1.1 (independent random variables are uncorrelated) that

$$\begin{aligned} \int_B E(\xi)\, dP &= E(\xi)E(1_B) \\ &= E(\xi 1_B) \\ &= \int_B \xi\, dP, \end{aligned}$$

which proves the assertion.

5) By Definition 2.4

$$\int_B E(\xi|\mathcal{G})\, dP = \int_B \xi\, dP$$

for every $B \in \mathcal{G}$, and

$$\int_B E(\xi|\mathcal{H})\, dP = \int_B \xi\, dP$$

for every $B \in \mathcal{H}$. Because $\mathcal{H} \subset \mathcal{G}$ it follows that

$$\int_B E(\xi|\mathcal{G})\, dP = \int_B E(\xi|\mathcal{H})\, dP$$

for every $B \in \mathcal{H}$. Applying Definition 2.4 once again, we obtain

$$E(E(\xi|\mathcal{G})|\mathcal{H}) = E(\xi|\mathcal{H}).$$

6) For any n we put

$$A_n = \left\{ E(\xi|\mathcal{G}) \le -\frac{1}{n} \right\}.$$

Then $A_n \in \mathcal{G}$. If $\xi \ge 0$ a.s., then

$$0 \le \int_{A_n} \xi\, dP = \int_{A_n} E(\xi|\mathcal{G})\, dP \le -\frac{1}{n} P(A_n),$$

which means that $P(A_n) = 0$. Because

$$\{E(\xi|\mathcal{G}) < 0\} = \bigcup_{n=1}^{\infty} A_n$$

it follows that

$$P\{E(\xi|\mathcal{G}) < 0\} = 0,$$

completing the proof. □

The next theorem, which will be stated without proof, involves the notion of a convex function, such as $\max(1, x)$ or e^{-x}, for example. In this course the theorem will be used mainly for $|x|$, which is also a convex function. In general, we call a function $\varphi : \mathbb{R} \to \mathbb{R}$ *convex* if for any $x, y \in \mathbb{R}$ and any $\lambda \in [0, 1]$

$$\varphi(\lambda x + (1 - \lambda) y) \leq \lambda\varphi(x) + (1 - \lambda)\varphi(y).$$

This condition means that the graph of φ lies below the cord connecting the points with coordinates $(x, \varphi(x))$ and $(y, \varphi(y))$.

Theorem 2.2 (Jensen's Inequality)

Let $\varphi : \mathbb{R} \to \mathbb{R}$ be a convex function and let ξ be an integrable random variable on a probability space $(\Omega, \mathcal{F}, P)$ such that $\varphi(\xi)$ is also integrable. Then

$$\varphi(E(\xi|\mathcal{G})) \leq E(\varphi(\xi)|\mathcal{G}) \quad \text{a.s.}$$

for any σ-field $\mathcal{G}$ on Ω contained in $\mathcal{F}$.

2.6 Various Exercises on Conditional Expectation

Exercise 2.13

Mrs. Jones has made a steak and kidney pie for her two sons. Eating more than a half of it will give indigestion to anyone. While she is away having tea with a neighbour, the older son helps himself to a piece of the pie. Then the younger son comes and has a piece of what is left by his brother. We assume that the size of each of the two pieces eaten by Mrs. Jones' sons is random and uniformly distributed over what is currently available. What is the expected size of the remaining piece given that neither son gets indigestion?

Hint All possible outcomes can be represented by pairs of numbers, the portions of the pie consumed by the two sons. Therefore Ω can be chosen as a subset of the plane. Observe that the older son is restricted only by the size of the pie, while the younger one is restricted by what is left by his brother. This will determine the shape of Ω. Next introduce a probability measure on Ω consistent with the conditions of the exercise. This can be done by means of a suitable density over Ω. Now you are in a position to compute the probability that neither son will get indigestion. What is the corresponding subset of Ω? Finally, define a random variable on Ω representing the portion of the pie left by the sons and compute the conditional expectation.

Exercise 2.14

As a probability space take $\Omega = [0,1)$ with the σ-field of Borel sets and the Lebesgue measure on $[0,1)$. Find $E(\xi|\eta)$ if

$$\xi(x) = 2x^2, \qquad \eta(x) = \begin{cases} 2x & \text{for } 0 \le x < \frac{1}{2}, \\ 2x-1 & \text{for } \frac{1}{2} \le x < 1. \end{cases}$$

Hint What do events in $\sigma(\eta)$ look like? What do $\sigma(\eta)$-measurable random variables look like? If you devise a neat way of describing these, it will make the task of finding $E(\xi|\eta)$ much easier. You will need to transform the integrals in condition 2) of Definition 2.3 to find a formula for the conditional expectation.

Exercise 2.15

Take $\Omega = [0,1]$ with the σ-field of Borel sets and P the Lebesgue measure on $[0,1]$. Let

$$\eta(x) = x(1-x)$$

for $x \in [0,1]$. Show that

$$E(\xi|\eta)(x) = \frac{\xi(x) + \xi(1-x)}{2}$$

for any $x \in [0,1]$.

Hint Observe that $\eta(x) = \eta(1-x)$. What does it tell you about the σ-field generated by η? Is $\frac{1}{2}(\xi(x) + \xi(1-x))$ measurable with respect to this σ-field? If so, it remains to verify condition 2) of Definition 2.3.

Exercise 2.16

Let ξ, η be integrable random variables with joint density $f_{\xi,\eta}(x,y)$. Show that

$$E(\xi|\eta) = \frac{\int_{\mathbb{R}} x\, f_{\xi,\eta}(x,\eta)\, dx}{\int_{\mathbb{R}} f_{\xi,\eta}(x,\eta)\, dx}.$$

Hint Study the solutions to Exercises 2.7 and 2.8.

Remark 2.3

If we put

$$f_{\xi,\eta}(x|y) = \frac{f_{\xi,\eta}(x,y)}{f_\eta(y)},$$

where

$$f_\eta(y) = \int_{\mathbb{R}} f_{\xi,\eta}(x,y)\,dx$$

is the density of η, then by the result in Exercise 2.16

$$E(\xi|\eta) = \int_{\mathbb{R}} x\, f_{\xi,\eta}(x|\eta)\,dx.$$

We call $f_{\xi,\eta}(x|y)$ the *conditional density* of ξ given η.

Exercise 2.17

Consider $L^2(\mathcal{F}) = L^2(\Omega, \mathcal{F}, P)$ as a Hilbert space with scalar product

$$L^2(\mathcal{F}) \times L^2(\mathcal{F}) \ni (\xi, \zeta) \mapsto E(\xi\zeta) \in \mathbb{R}.$$

Show that if ξ is a random variable in $L^2(\mathcal{F})$ and $\mathcal{G}$ is a σ-field contained in $\mathcal{F}$, then $E(\xi|\mathcal{G})$ is the orthogonal projection of ξ onto the subspace $L^2(\mathcal{G})$ in $L^2(\mathcal{F})$ consisting of $\mathcal{G}$-measurable random variables.

Hint Observe that condition 2) of Definition 2.4 means that $\xi - E(\xi|\mathcal{G})$ is orthogonal (in the sense of the scalar product above) to the indicator function 1_A for any $A \in \mathcal{G}$.

2.7 Solutions

Solution 2.1

Since $P(\Omega) = 1$ and $\int_\Omega \xi\,dP = E(\xi)$,

$$E(\xi|\Omega) = \frac{1}{P(\Omega)} \int_\Omega \xi\,dP = E(\xi).$$

Solution 2.2

By Definition 2.1

$$\begin{aligned} E(1_A|B) &= \frac{1}{P(B)} \int_B 1_A\,dP \\ &= \frac{1}{P(B)} \int_{A\cap B} dP \end{aligned}$$

$$= \frac{P(A \cap B)}{P(B)}$$
$$= P(A|B).$$

Solution 2.3

Since η is constant, it has only one value $c \in \mathbb{R}$, for which

$$\{\eta = c\} = \Omega.$$

Therefore $E(\xi|\eta)$ is constant and equal to

$$E(\xi|\eta)(\omega) = E(\xi|\{\eta = c\}) = E(\xi|\Omega) = E(\xi)$$

for each $\omega \in \Omega$. The last equality has been verified in Exercise 2.1.

Solution 2.4

The indicator function 1_B takes two values 1 and 0. The sets on which these values are taken are

$$\{1_B = 1\} = B, \qquad \{1_B = 0\} = \Omega \setminus B.$$

Thus, for $\omega \in B$

$$E(1_A|1_B)(\omega) = E(1_A|B) = P(A|B),$$

as in Exercise 2.2. Similarly, for $\omega \in \Omega \setminus B$

$$E(1_A|1_B)(\omega) = E(1_A|\Omega \setminus B) = P(A|\Omega \setminus B).$$

Solution 2.5

First we observe that

$$\int_B E(\xi|B)\, dP = \int_B \left(\frac{1}{P(B)} \int_B \xi\, dP \right) dP = \int_B \xi\, dP \tag{2.4}$$

for any event B.

Since η is discrete, it has countably many values $y_1, y_2, \ldots$. We can assume that these values are pairwise distinct, i.e. $y_i \neq y_j$ if $i \neq j$. The sets $\{\eta = y_1\}, \{\eta = y_2\}, \ldots$ are then pairwise disjoint and they cover the whole space Ω. Therefore, by (2.4)

$$\begin{aligned} E(E(\xi|\eta)) &= \int_\Omega E(\xi|\eta)\, dP \\ &= \sum_n \int_{\{\eta = y_n\}} E(\xi|\{\eta = y_n\})\, dP \end{aligned}$$

$$
\begin{aligned}
&= \sum_n \int_{\{\eta=y_n\}} \xi \, dP \\
&= \int_\Omega \xi \, dP \\
&= E(\xi).
\end{aligned}
$$

Solution 2.6

First we need to describe the σ-field $\sigma(\eta)$ generated by η. Observe that η is symmetric about $\frac{1}{2}$,

$$\eta(x) = \eta(1-x)$$

for any $x \in [0,1]$. We claim that $\sigma(\eta)$ consists of all Borel sets $A \subset [0,1]$ symmetric about $\frac{1}{2}$, i.e. such that

$$A = 1 - A. \tag{2.5}$$

Indeed, if A is such a set, then

$$A = \{\eta \in 2A - 1\} \in \sigma(\eta).$$

On the other hand, if $A \in \sigma(\eta)$, then there is a Borel set B in $\mathbb{R}$ such that $A = \{\eta \in B\}$. Then

$$
\begin{aligned}
x \in A &\Leftrightarrow \eta(x) \in B \\
&\Leftrightarrow \eta(1-x) \in B \\
&\Leftrightarrow 1 - x \in A,
\end{aligned}
$$

so A satisfies (2.5).

We are ready to find $E(\xi|\eta)$. If it is to be $\sigma(\eta)$-measurable, it must be symmetric about $\frac{1}{2}$, i.e.

$$E(\xi|\eta)(x) = E(\xi|\eta)(1-x)$$

for any $x \in [0,1]$. For any $A \in \sigma(\eta)$ we shall transform the integral below so as to make the integrand symmetric about $\frac{1}{2}$:

$$
\begin{aligned}
\int_A 2x^2 \, dx &= \int_A x^2 \, dx + \int_A x^2 \, dx \\
&= \int_A x^2 \, dx + \int_{1-A} (1-x)^2 \, dx \\
&= \int_A x^2 \, dx + \int_A (1-x)^2 \, dx \\
&= \int_A \left(x^2 + (1-x)^2\right) dx.
\end{aligned}
$$

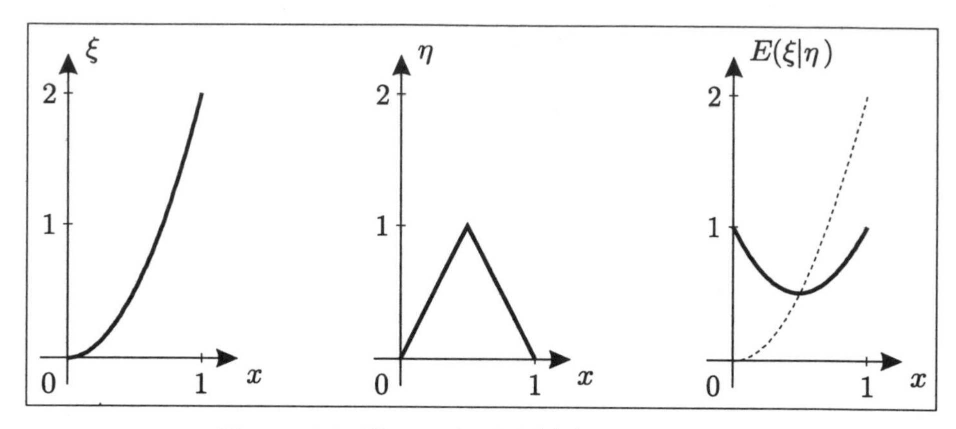

Figure 2.3. The graph of $E(\xi|\eta)$ in Exercise 2.6

It follows that

$$E(\xi|\eta)(x) = x^2 + (1-x)^2.$$

The graphs of ξ, η and $E(\xi|\eta)$ are shown in Figure 2.3.

Solution 2.7

Since

$$\{\eta \in B\} = [0,1] \times B$$

for any Borel set B, we have

$$\begin{aligned}
\int_{\{\eta\in B\}} \xi \, dP &= \int_B \int_{\mathbb{R}} x \, f_{\xi,\eta}(x,y) \, dx \, dy \\
&= \int_B \left(\int_{[0,1]} x(x+y) \, dx \right) dy \\
&= \int_B \left(\frac{1}{3} + \frac{1}{2} y \right) dy
\end{aligned}$$

and

$$\begin{aligned}
\int_{\{\eta\in B\}} \frac{2+3\eta}{3+6\eta} \, dP &= \int_B \int_{\mathbb{R}} \frac{2+3y}{3+6y} f_{\xi,\eta}(x,y) \, dx \, dy \\
&= \int_B \frac{2+3y}{3+6y} \left(\int_{[0,1]} (x+y) \, dx \right) dy \\
&= \int_B \left(\frac{1}{3} + \frac{1}{2} y \right) dy.
\end{aligned}$$

Because each event in $\sigma(\eta)$ is of the form $\{\eta \in B\}$ for some Borel set B, this means that condition 2) of Definition 2.3 is satisfied. The random variable $\frac{2+3\eta}{3+6\eta}$

is $\sigma(\eta)$-measurable, so condition 1) holds too. It follows that

$$E(\xi|\eta) = \frac{2+3\eta}{3+6\eta}.$$

Solution 2.8

We are looking for a Borel function $F(y)$ such that for any Borel set B

$$\int_{\{\eta\in B\}} \xi\, dP = \int_{\{\eta\in B\}} F(\eta)\, dP. \tag{2.6}$$

Then $E(\xi|\eta) = F(\eta)$ by Definition 2.3.

We shall transform both integrals above using the joint density $f_{\xi,\eta}(x,y)$ in much the same way as in the solution to Exercise 2.7, except that here we do not know the exact form of $F(y)$. Namely,

$$\begin{aligned}\int_{\{\eta\in B\}} \xi\, dP &= \int_B \int_{\mathbb{R}} x\, f_{\xi,\eta}(x,y)\, dx\, dy \\ &= \frac{3}{2}\int_B \left(\int_{[0,1]} x\left(x^2+y^2\right) dx\right) dy \\ &= \frac{3}{2}\int_B \left(\frac{1}{4}+\frac{1}{2}y^2\right) dy\end{aligned}$$

and

$$\begin{aligned}\int_{\{\eta\in B\}} F(\eta)\, dP &= \int_B \int_{\mathbb{R}} F(y)\, f_{\xi,\eta}(x,y)\, dx\, dy \\ &= \frac{3}{2}\int_B F(y)\left(\int_{[0,1]} \left(x^2+y^2\right) dx\right) dy \\ &= \frac{3}{2}\int_B F(y)\left(\frac{1}{3}+y^2\right) dy.\end{aligned}$$

Then, (2.6) will hold for any Borel set B if

$$F(y) = \frac{\frac{1}{4}+\frac{1}{2}y^2}{\frac{1}{3}+y^2} = \frac{3+6y^2}{4+12y^2}.$$

It follows that

$$E(\xi|\eta) = F(\eta) = \frac{3+6\eta^2}{4+12\eta^2}.$$

Solution 2.9

We are looking for a Borel function $F(y)$ such that for any Borel set $B \subset \mathbb{R}$

$$\int_{\{\eta\in B\}} \xi^2\, dP = \int_{\{\eta\in B\}} F(\eta)\, dP. \tag{2.7}$$

Then, by Definition 2.3 we shall have $E(\xi^2|\eta) = F(\eta)$.

Let us transform both sides of (2.7). To do so we observe that the random variables ξ and η have uniform joint distribution over the unit disc $\Omega = \{(x,y) : x^2 + y^2 \le 1\}$, with density

$$f_{\xi,\eta}(x,y) = \frac{1}{\pi}$$

if $x^2 + y^2 \le 1$, and $f_{\xi,\eta}(x,y) = 0$ otherwise. It follows that

$$\begin{aligned}\int_{\{\eta\in B\}} \xi^2 \, dP &= \int_B \int_{\mathbb{R}} x^2 f_{\xi,\eta}(x,y)\, dx\, dy \\ &= \frac{1}{\pi}\int_B \int_{-\sqrt{1-y^2}}^{\sqrt{1-y^2}} x^2 \, dx\, dy \\ &= \frac{2}{3\pi}\int_B \left(1-y^2\right)^{3/2} dy\end{aligned}$$

and

$$\begin{aligned}\int_{\{\eta\in B\}} F(\eta)\, dP &= \int_B \int_{\mathbb{R}} F(y)\, f_{\xi,\eta}(x,y)\, dx\, dy \\ &= \frac{1}{\pi}\int_B F(y) \int_{-\sqrt{1-y^2}}^{\sqrt{1-y^2}} dx\, dy \\ &= \frac{2}{\pi}\int_B F(y)\left(1-y^2\right)^{1/2} dy.\end{aligned}$$

If (2.7) is to be satisfied for all Borel sets B, then

$$F(y) = \frac{1}{3}\left(1-y^2\right).$$

This means that

$$E\left(\xi^2|\eta\right)(x,y) = F(\eta(x,y)) = F(y) = \frac{1}{3}\left(1-y^2\right)$$

for any (x,y) in Ω.

Solution 2.10

If $\mathcal{G} = \{\emptyset, \Omega\}$, then any constant random variable is $\mathcal{G}$-measurable. Since

$$\int_\Omega \xi\, dP = E(\xi) = \int_\Omega E(\xi)\, dP$$

and

$$\int_\emptyset \xi\, dP = 0 = \int_\emptyset E(\xi)\, dP,$$

it follows that $E(\xi|\mathcal{G}) = E(\xi)$ a.s., as required.

Solution 2.11

Because the trivial identity

$$\int_A \xi \, dP = \int_A \xi \, dP$$

holds for any $A \in \mathcal{G}$ and ξ is $\mathcal{G}$-measurable, it follows that $E(\xi|\mathcal{G}) = \xi$ a.s.

Solution 2.12

By Definition 2.3

$$\int_B E(\xi|\mathcal{G}) \, dP = \int_B \xi \, dP,$$

since $B \in \mathcal{G}$. It follows that

$$\begin{aligned} E(E(\xi|\mathcal{G})|B) &= \frac{1}{P(B)} \int_B E(\xi|\mathcal{G}) \, dP \\ &= \frac{1}{P(B)} \int_B \xi \, dP \\ &= E(\xi|B). \end{aligned}$$

Solution 2.13

The whole pie will be represented by the interval $[0,1]$. Let $x \in [0,1]$ be the portion consumed by the older son. Then $[0, 1-x]$ will be available to the younger one, who takes a portion of size $y \in [0, 1-x]$. The set of all possible outcomes is

$$\Omega = \{(x,y) : x, y \geq 0, x + y \leq 1\}.$$

The event that neither of Mrs. Jones' sons will get indigestion is

$$A = \left\{(x,y) \in \Omega : x, y < \frac{1}{2}\right\}.$$

These sets are shown in Figure 2.4. If x is uniformly distributed over $[0,1]$ and y is uniformly distributed over $[0, 1-x]$, then the probability measure P over Ω with density

$$f(x,y) = \frac{1}{1-x}$$

will describe the joint distribution of outcomes (x,y), see Figure 2.5.

Now we are in a position to compute

$$\begin{aligned} P(A) &= \int_A f(x,y) \, dx \, dy \\ &= \int_0^{\frac{1}{2}} \int_0^{\frac{1}{2}} \frac{1}{1-x} \, dx \, dy \\ &= \ln \sqrt{2}. \end{aligned}$$

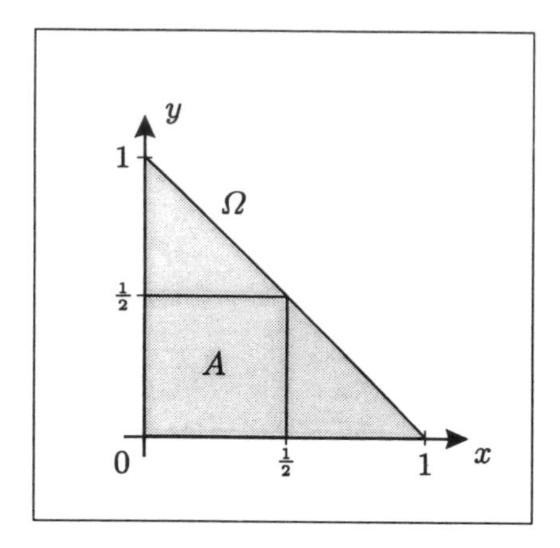

Figure 2.4. The sets Ω and A in Exercise 2.13

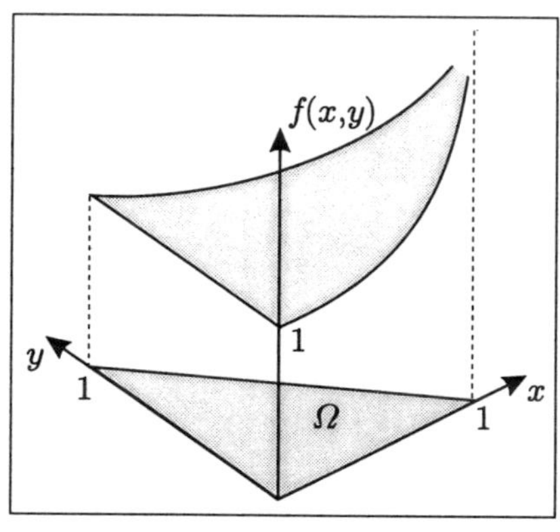

Figure 2.5. The density $f(x, y)$ in Exercise 2.13

The random variable

$$\xi(x, y) = 1 - x - y$$

defined on Ω represents the size of the portion left by Mrs. Jones' sons. Finally, we find that

$$\begin{aligned} E(\xi|A) &= \frac{1}{P(A)} \int_A (1 - x - y)\, f(x, y)\, dx\, dy \\ &= \frac{1}{\ln\sqrt{2}} \int_0^{\frac{1}{2}} \int_0^{\frac{1}{2}} \frac{1 - x - y}{1 - x}\, dx\, dy \\ &= \frac{1 - \ln\sqrt{2}}{\ln 4}. \end{aligned}$$

Solution 2.14

The σ-field $\sigma(\eta)$ generated by η consists of sets of the form $B \cup \left(B + \frac{1}{2}\right)$ for some Borel set $B \subset [0, \frac{1}{2})$. Thus, we are looking for a $\sigma(\eta)$-measurable random variable ζ such that for each Borel set $B \subset [0, \frac{1}{2})$

$$\int_{B \cup (B+\frac{1}{2})} \xi(x)\, dx = \int_{B \cup (B+\frac{1}{2})} \zeta(x)\, dx. \tag{2.8}$$

Then $E(\xi|\eta) = \zeta$ by Definition 2.3.

Transforming the integral on the left-hand side, we obtain

$$\begin{aligned}
\int_{B \cup (B+\frac{1}{2})} \xi(x)\, dx &= \int_B 2x^2\, dx + \int_{B+\frac{1}{2}} 2x^2\, dx \\
&= \int_B 2x^2\, dx + \int_B 2\left(x+\tfrac{1}{2}\right)^2 dx \\
&= 2\int_B \left(x^2 + \left(x+\tfrac{1}{2}\right)^2\right) dx.
\end{aligned}$$

For ζ to be $\sigma(\eta)$-measurable it must satisfy

$$\zeta(x) = \zeta\left(x + \tfrac{1}{2}\right) \tag{2.9}$$

for each $x \in [0, \frac{1}{2})$. Then

$$\begin{aligned}
\int_{B \cup (B+\frac{1}{2})} \zeta(x)\, dP &= \int_B \zeta(x)\, dx + \int_{B+\frac{1}{2}} \zeta(x)\, dx \\
&= \int_B \zeta(x)\, dx + \int_B \zeta\left(x+\tfrac{1}{2}\right) dx \\
&= \int_B \zeta(x)\, dx + \int_B \zeta(x)\, dx \\
&= 2\int_B \zeta(x)\, dx.
\end{aligned}$$

If (2.8) is to hold for any Borel set $B \subset [0, \frac{1}{2})$, then

$$\zeta(x) = x^2 + \left(x + \tfrac{1}{2}\right)^2$$

for each $x \in [0, \frac{1}{2})$. The values of $\zeta(x)$ for $x \in [\frac{1}{2}, 1)$ can be obtained from (2.9). It follows that

$$E(\xi|\eta)(x) = \zeta(x) = \begin{cases} x^2 + \left(x+\frac{1}{2}\right)^2 & \text{for } 0 \le x < \frac{1}{2}, \\ \left(x-\frac{1}{2}\right)^2 + x^2 & \text{for } \frac{1}{2} \le x < 1. \end{cases}$$

The graphs of ξ, η and $E(\xi|\eta)$ are shown in Figure 2.6.

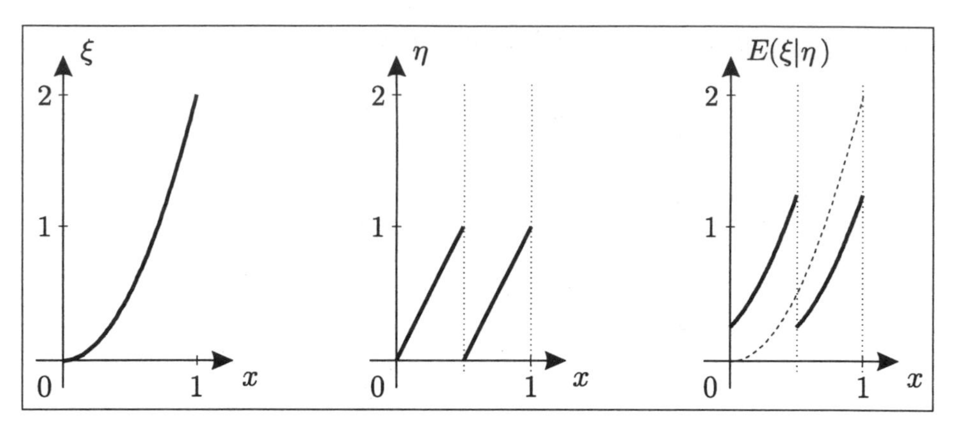

Figure 2.6. The graph of $E(\xi|\eta)$ in Exercise 2.14

Solution 2.15

Since $\eta(x) = \eta(1-x)$, the σ-field $\sigma(\eta)$ consists of Borel sets $B \subset [0,1]$ such that

$$B = 1 - B,$$

where $1 - B = \{1 - x : x \in B\}$. For any such B

$$\begin{aligned}
\int_B \xi(x)\,dx &= \frac{1}{2}\int_B \xi(x)\,dx + \frac{1}{2}\int_B \xi(x)\,dx \\
&= \frac{1}{2}\int_B \xi(x)\,dx + \frac{1}{2}\int_{1-B} \xi(1-x)\,dx \\
&= \frac{1}{2}\int_B \xi(x)\,dx + \frac{1}{2}\int_B \xi(1-x)\,dx \\
&= \int_B \frac{\xi(x)+\xi(1-x)}{2}\,dx.
\end{aligned}$$

Because $\frac{1}{2}(\xi(x) + \xi(1-x))$ is $\sigma(\eta)$-measurable, it follows that

$$E(\xi|\eta)(x) = \frac{\xi(x)+\xi(1-x)}{2}.$$

Solution 2.16

We are looking for a Borel function $F(y)$ such that

$$\int_{\{\eta\in B\}} \xi\,dP = \int_{\{\eta\in B\}} F(\eta)\,dP$$

for any Borel set B in $\mathbb{R}$. Because $F(\eta)$ is $\sigma(\eta)$-measurable and each event in $\sigma(\eta)$ can be written as $\{\eta \in B\}$ for some Borel set B, this will mean that $E(\xi|\eta) = F(\eta)$.

Let us transform the two integrals above using the joint density of ξ and η:

$$\begin{aligned}\int_{\{\eta\in B\}} \xi\, dP &= \int_B \int_{\mathbb{R}} x\, f_{\xi,\eta}(x,y)\, dx\, dy \\ &= \int_B \left(\int_{\mathbb{R}} x\, f_{\xi,\eta}(x,y)\, dx \right) dy\end{aligned}$$

and

$$\begin{aligned}\int_{\{\eta\in B\}} F(\eta)\, dP &= \int_B \int_{\mathbb{R}} F(y)\, f_{\xi,\eta}(x,y)\, dx\, dy \\ &= \int_B F(y) \left(\int_{\mathbb{R}} f_{\xi,\eta}(x,y)\, dx \right) dy.\end{aligned}$$

If these two integrals are to be equal for each Borel set B, then

$$F(y) = \frac{\int_{\mathbb{R}} x\, f_{\xi,\eta}(x,y)\, dx}{\int_{\mathbb{R}} f_{\xi,\eta}(x,y)\, dx}.$$

It follows that

$$E(\xi|\eta) = F(\eta) = \frac{\int_{\mathbb{R}} x\, f_{\xi,\eta}(x,\eta)\, dx}{\int_{\mathbb{R}} f_{\xi,\eta}(x,\eta)\, dx}.$$

Solution 2.17

We denote by ζ the orthogonal projection of ξ onto the subspace $L^2(\mathcal{G}) \subset L^2(\mathcal{F})$ consisting of $\mathcal{G}$-measurable random variables. Thus, $\xi - \zeta$ is orthogonal to $L^2(\mathcal{G})$, that is,

$$E[(\xi - \zeta)\gamma] = 0$$

for each $\gamma \in L^2(\mathcal{G})$. But for any $A \in \mathcal{G}$ the indicator function 1_A belongs to $L^2(\mathcal{G})$, so

$$E[(\xi - \zeta)1_A] = 0.$$

Therefore

$$\int_A \xi\, dP = E(\xi 1_A) = E(\zeta 1_A) = \int_A \zeta\, dP$$

for any $A \in \mathcal{G}$. This means that $\zeta = E(\xi|\mathcal{G})$.

3
Martingales in Discrete Time

3.1 Sequences of Random Variables

A sequence $\xi_1, \xi_2, \ldots$ of random variables is typically used as a mathematical model of the outcomes of a series of random phenomena, such as coin tosses or the value of the FTSE All-Share Index at the London Stock Exchange on consecutive business days. The random variables in such a sequence are indexed by whole numbers, which are customarily referred to as *discrete time*. It is important to understand that these whole numbers are not necessarily related to the physical time when the events modelled by the sequence actually occur. Discrete time is used to keep track of the order of events, which may or may not be evenly spaced in physical time. For example, the share index is recorded only on business days, but not on Saturdays, Sundays or any other holidays. Rather than tossing a coin repeatedly, we may as well toss 100 coins at a time and count the outcomes.

Definition 3.1

The sequence of numbers $\xi_1(\omega), \xi_2(\omega), \ldots$ for any fixed $\omega \in \Omega$ is called a *sample path.*

A sample path for a sequence of coin tosses is presented in Figure 3.1 ($+1$ stands for heads and -1 for tails). Figure 3.2 shows the sample path of the FTSE All-Share Index up to 1997. Strictly speaking the pictures should con-

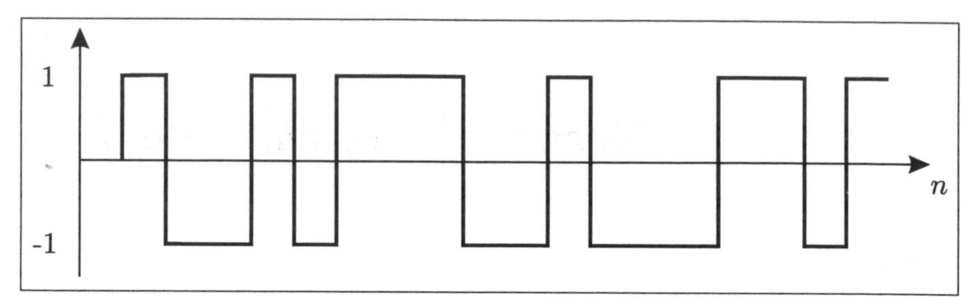

Figure 3.1. Sample path for a sequence of coin tosses

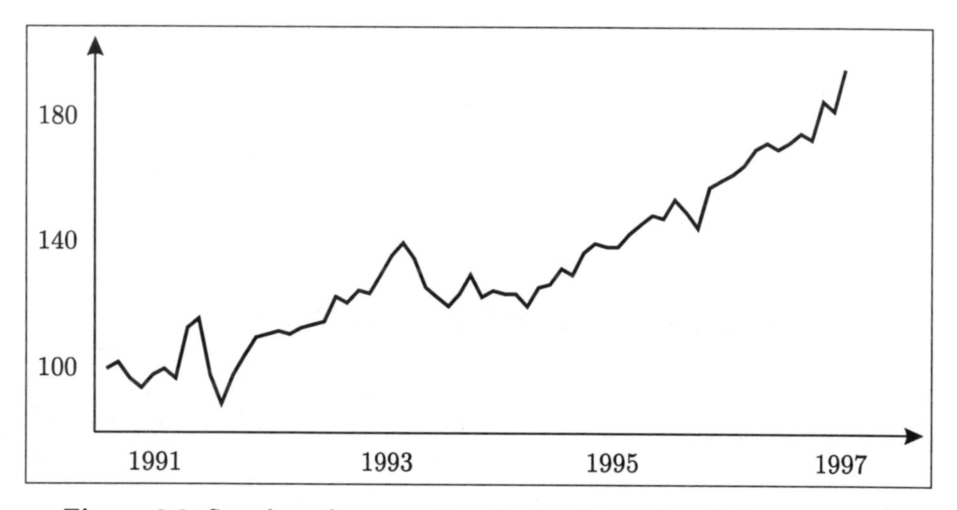

Figure 3.2. Sample path representing the FTSE All-Share Index up to 1997

sist of dots, representing the values $\xi_1(\omega), \xi_2(\omega), \ldots$, but it is customary to connect them by a broken line for illustration purposes.

3.2 Filtrations

As the time n increases, so does our knowledge about what has happened in the past. This can be modelled by a filtration as defined below.

Definition 3.2

A sequence of σ-fields $\mathcal{F}_1, \mathcal{F}_2, \ldots$ on Ω such that

$$\mathcal{F}_1 \subset \mathcal{F}_2 \subset \cdots \subset \mathcal{F}$$

is called a *filtration.*

Here $\mathcal{F}_n$ represents our knowledge at time n. It contains all events A such that at time n it is possible to decide whether A has occurred or not. As n increases, there will be more such events A, i.e. the family $\mathcal{F}_n$ representing our knowledge will become larger. (The longer you live the wiser you become!)

Example 3.1

For a sequence $\xi_1, \xi_2, \ldots$ of coin tosses we take $\mathcal{F}_n$ to be the σ-field generated by $\xi_1, \ldots, \xi_n$,

$$\mathcal{F}_n = \sigma(\xi_1, \ldots, \xi_n).$$

Let

$$A = \{\text{the first 5 tosses produce at least 2 heads}\}.$$

At discrete time $n = 5$, i.e. once the coin has been tossed five times, it will be possible to decide whether A has occurred or not. This means that $A \in \mathcal{F}_5$. However, at $n = 4$ it is not always possible to tell if A has occurred or not. If the outcomes of the first four tosses are, say,

tails, tails, heads, tails,

then the event A remains undecided. We will have to toss the coin once more to see what happens. Therefore $A \notin \mathcal{F}_4$.

This example illustrates another relevant issue. Suppose that the outcomes of the first four coin tosses are

tails, heads, tails, heads.

In this case it is possible to tell that A has occurred already at $n = 4$, whatever the outcome of the fifth toss will be. It does not mean, however, that A belongs to $\mathcal{F}_4$. The point is that for A to belong to $\mathcal{F}_4$ it must be possible to tell whether A has occurred or not after the first four tosses, *no matter what the first four outcomes are.* This is clearly not so in the example in hand.

Exercise 3.1

Let $\xi_1, \xi_2, \ldots$ be a sequence of coin tosses and let $\mathcal{F}_n$ be the σ-field generated by $\xi_1, \ldots, \xi_n$. For each of the following events find the smallest n such that the event belongs to $\mathcal{F}_n$:
$A = \{$the first occurrence of heads is preceeded by no more than 10 tails$\}$,
$B = \{$there is at least 1 head in the sequence $\xi_1, \xi_2, \ldots\}$,
$C = \{$the first 100 tosses produce the same outcome$\}$,
$D = \{$there are no more than 2 heads and 2 tails among the first 5 tosses$\}$.

Hint To find the smallest element in a set of numbers you need to make sure that the set is non-empty in the first place.

Suppose that $\xi_1, \xi_2, \ldots$ is a sequence of random variables and $\mathcal{F}_1, \mathcal{F}_2, \ldots$ is a filtration. A priori, they may have nothing in common. However, in practice the filtration will usually contain the knowledge accumulated by observing the outcomes of the random sequence, as in Example 3.1. The condition in the definition below means that $\mathcal{F}_n$ contains everything that can be learned from the values of $\xi_1, \ldots, \xi_n$. In general, it may and often does contain more information.

Definition 3.3

We say that a sequence of random variables $\xi_1, \xi_2, \ldots$ is *adapted* to a filtration $\mathcal{F}_1, \mathcal{F}_2, \ldots$ if ξ_n is $\mathcal{F}_n$-measurable for each $n = 1, 2, \ldots$.

Example 3.2

If $\mathcal{F}_n = \sigma(\xi_1, \ldots, \xi_n)$ is the σ-field generated by $\xi_1, \ldots, \xi_n$, then $\xi_1, \xi_2, \ldots$ is adapted to $\mathcal{F}_1, \mathcal{F}_2, \ldots$.

Exercise 3.2

Show that

$$\mathcal{F}_n = \sigma(\xi_1, \ldots, \xi_n)$$

is the smallest filtration such that the sequence $\xi_1, \xi_2, \ldots$ is adapted to $\mathcal{F}_1$, $\mathcal{F}_2, \ldots$. That is to say, if $\mathcal{G}_1, \mathcal{G}_2, \ldots$ is another filtration such that $\xi_1, \xi_2, \ldots$ is adapted to $\mathcal{G}_1, \mathcal{G}_2, \ldots$, then $\mathcal{F}_n \subset \mathcal{G}_n$ for each n.

Hint For $\sigma(\xi_1, \ldots, \xi_n)$ to be contained in $\mathcal{G}_n$ you need to show that $\xi_1, \ldots, \xi_n$ are $\mathcal{G}_n$-measurable.

3.3 Martingales

The concept of a martingale has its origin in gambling, namely, it describes a fair game of chance, which will be discussed in detail in the next section. Similarly, the notions of submartingale and supermartingale defined below are related to favourable and unfavourable games of chance. Some aspects of gambling are inherent in the mathematics of finance, in particular, the theory of financial derivatives such as options. Not surprisingly, martingales play a crucial role there. In fact, martingales reach well beyond game theory and appear in various areas of modern probability and stochastic analysis, notably, in diffusion theory. First of all, let us introduce the basic definitions and properties in the case of discrete time.

Definition 3.4

A sequence $\xi_1, \xi_2, \ldots$ of random variables is called a *martingale* with respect to a filtration $\mathcal{F}_1, \mathcal{F}_2, \ldots$ if

1) ξ_n is integrable for each $n = 1, 2, \ldots$;
2) $\xi_1, \xi_2, \ldots$ is adapted to $\mathcal{F}_1, \mathcal{F}_2, \ldots$;
3) $E(\xi_{n+1}|\mathcal{F}_n) = \xi_n$ a.s. for each $n = 1, 2, \ldots$.

Example 3.3

Let $\eta_1, \eta_2, \ldots$ be a sequence of independent integrable random variables such that $E(\eta_n) = 0$ for all $n = 1, 2, \ldots$. We put

$$\begin{aligned} \xi_n &= \eta_1 + \cdots + \eta_n, \\ \mathcal{F}_n &= \sigma(\eta_1, \ldots, \eta_n). \end{aligned}$$

Then ξ_n is adapted to the filtration $\mathcal{F}_n$, and it is integrable because

$$\begin{aligned} E(|\xi_n|) &= E(|\eta_1 + \cdots + \eta_n|) \\ &\leq E(|\eta_1|) + \cdots + E(|\eta_n|) \\ &< \infty. \end{aligned}$$

Moreover,

$$\begin{aligned} E(\xi_{n+1}|\mathcal{F}_n) &= E(\eta_{n+1}|\mathcal{F}_n) + E(\xi_n|\mathcal{F}_n) \\ &= E(\eta_{n+1}) + \xi_n \\ &= \xi_n, \end{aligned}$$

since η_{n+1} is independent of $\mathcal{F}_n$ ('and independent condition drops out') and ξ_n is $\mathcal{F}_n$-measurable ('taking out what is known'). This means that ξ_n is a martingale with respect to $\mathcal{F}_n$.

Example 3.4

Let ξ be an integrable random variable and let $\mathcal{F}_1, \mathcal{F}_2, \ldots$ be a filtration. We put

$$\xi_n = E(\xi|\mathcal{F}_n)$$

for $n = 1, 2, \ldots$.

Then ξ_n is $\mathcal{F}_n$-measurable,

$$|\xi_n| = |E(\xi|\mathcal{F}_n)| \leq E(|\xi|\ |\mathcal{F}_n),$$

which implies that

$$E(|\xi_n|) \leq E(E(|\xi|\ |\mathcal{F}_n)) = E(|\xi|) < \infty,$$

and

$$E(\xi_{n+1}|\mathcal{F}_n) = E(E(\xi|\mathcal{F}_{n+1})|\mathcal{F}_n) = E(\xi|\mathcal{F}_n) = \xi_n,$$

since $\mathcal{F}_n \subset \mathcal{F}_{n+1}$ (the tower property of conditional expectation). Therefore ξ_n is a martingale with respect to $\mathcal{F}_n$.

Exercise 3.3

Show that if ξ_n is a martingale with respect to $\mathcal{F}_n$, then

$$E(\xi_1) = E(\xi_2) = \cdots \ .$$

Hint What is the expectation of $E(\xi_{n+1}|\mathcal{F}_n)$?

Exercise 3.4

Suppose that ξ_n is a martingale with respect to a filtration $\mathcal{F}_n$. Show that ξ_n is a martingale with respect to the filtration

$$\mathcal{G}_n = \sigma(\xi_1, \ldots, \xi_n).$$

Hint Observe that $\mathcal{G}_n \subset \mathcal{F}_n$ and use the tower property of conditional expectation.

Exercise 3.5

Let ξ_n be a symmetric random walk, that is,

$$\xi_n = \eta_1 + \cdots + \eta_n,$$

where $\eta_1, \eta_2, \ldots$ is a sequence of independent identically distributed random variables such that

$$P\{\eta_n = 1\} = P\{\eta_n = -1\} = \frac{1}{2}$$

(a sequence of coin tosses, for example). Show that $\xi_n^2 - n$ is a martingale with respect to the filtration

$$\mathcal{F}_n = \sigma(\eta_1, \ldots, \eta_n).$$

Hint You want to transform $E\left(\xi_{n+1}^2-(n+1)\,|\mathcal{F}_n\right)$ to obtain ξ_n^2-n. Write

$$\begin{aligned}\xi_{n+1}^2 &= (\xi_n+\eta_{n+1})^2\\ &= \eta_{n+1}^2+2\eta_{n+1}\xi_n+\xi_n^2\end{aligned}$$

and observe that ξ_n is $\mathcal{F}_n$-measurable, while η_{n+1} is independent of $\mathcal{F}_n$. To transform the conditional expectation you can 'take out what is known' and use the fact that 'an independent condition drops out'. Do not forget to verify that ξ_n^2-n is integrable and adapted to $\mathcal{F}_n$.

Exercise 3.6

Let ξ_n be a symmetric random walk and $\mathcal{F}_n$ the filtration defined in Exercise 3.5. Show that

$$\zeta_n=(-1)^n\cos(\pi\xi_n)$$

is a martingale with respect to $\mathcal{F}_n$.

Hint You want to transform $E((-1)^{n+1}\cos(\pi\xi_{n+1})|\mathcal{F}_n)$ to obtain $(-1)^n\cos(\pi\xi_n)$. Use a similar argument as in Exercise 3.5 to achieve this. But, first of all, make sure that ζ_n is integrable and adapted to $\mathcal{F}_n$.

Definition 3.5

We say that $\xi_1,\xi_2,\dots$ is a *supermartingale* (*submartingale*) with respect to a filtration $\mathcal{F}_1,\mathcal{F}_2,\dots$ if

1) ξ_n is integrable for each $n=1,2,\dots$;
2) $\xi_1,\xi_2,\dots$ is adapted to $\mathcal{F}_1,\mathcal{F}_2,\dots$;
3) $E(\xi_{n+1}|\mathcal{F}_n)\le\xi_n$ (respectively, $E(\xi_{n+1}|\mathcal{F}_n)\ge\xi_n$) a.s. for each $n=1,2,\dots$.

Exercise 3.7

Let ξ_n be a sequence of square integrable random variables. Show that if ξ_n is a martingale with respect to a filtration $\mathcal{F}_n$, then ξ_n^2 is a submartingale with respect to the same filtration.

Hint Use Jensen's inequality with convex function $\varphi(x)=x^2$.

3.4 Games of Chance

Suppose that you take part in a game such as the roulette, for example. Let $\eta_1,\eta_2,\dots$ be a sequence of integrable random variables, where η_n are your *winnings (or losses) per unit stake* in game n. If your stake in each game is one, then your total winnings after n games will be

$$\xi_n = \eta_1 + \cdots + \eta_n.$$

We take the filtration

$$\mathcal{F}_n = \sigma(\eta_1, \ldots, \eta_n)$$

and also put $\xi_0 = 0$ and $\mathcal{F}_0 = \{\emptyset, \Omega\}$ for notational simplicity.

If $n-1$ rounds of the game have been played so far, your accumulated knowledge will be represented by the σ-field $\mathcal{F}_{n-1}$. The game is fair if

$$E(\xi_n|\mathcal{F}_{n-1}) = \xi_{n-1},$$

that is, you expect that your fortune at step n will on average be the same as at step $n-1$. The game will be favourable to you if

$$E(\xi_n|\mathcal{F}_{n-1}) \geq \xi_{n-1},$$

and unfavourable to you if

$$E(\xi_n|\mathcal{F}_{n-1}) \leq \xi_{n-1}$$

for $n = 1, 2, \ldots$. This corresponds to ξ_n being, respectively, a martingale, a submartingale, or a supermartingale with respect to $\mathcal{F}_n$, see Definitions 3.4 and 3.5.

Suppose that you can vary the stake to be α_n in game n. (In particular, α_n may be zero if you refrain from playing the nth game; it may even be negative if you own the casino and can accept other people's bets.) When the time comes to decide your stake α_n, you will know the outcomes of the first $n-1$ games. Therefore it is reasonable to assume that α_n is $\mathcal{F}_{n-1}$-measurable, where $\mathcal{F}_{n-1}$ represents your knowledge accumulated up to and including game $n-1$. In particular, since nothing is known before the first game, we take $\mathcal{F}_0 = \{\emptyset, \Omega\}$.

Definition 3.6

A *gambling strategy* $\alpha_1, \alpha_2, \ldots$ (with respect to a filtration $\mathcal{F}_1, \mathcal{F}_2, \ldots$) is a sequence of random variables such that α_n is $\mathcal{F}_{n-1}$-measurable for each $n = 1, 2, \ldots$, where $\mathcal{F}_0 = \{\emptyset, \Omega\}$. (Outside the context of gambling such a sequence of random variables α_n is called *previsible*.)

If you follow a strategy $\alpha_1, \alpha_2, \ldots$, then your *total winnings* after n games will be

$$\begin{aligned}\zeta_n &= \alpha_1\eta_1 + \cdots + \alpha_n\eta_n \\ &= \alpha_1(\xi_1 - \xi_0) + \cdots + \alpha_n(\xi_n - \xi_{n-1}).\end{aligned}$$

We also put $\zeta_0 = 0$ for convenience.

The following proposition has important consequences for gamblers. It means that a fair game will always turn into a fair one, no matter which gambling strategy is used. If one is not in a position to wager negative sums of money (e.g. to run a casino), it will be impossible to turn an unfavourable game into a favourable one or vice versa. You cannot beat the system! The boundedness of the sequence α_n means that your available capital is bounded and so is your credit limit.

Proposition 3.1

Let $\alpha_1, \alpha_2, \ldots$ be a gambling strategy.

1) If $\alpha_1, \alpha_2, \ldots$ is a bounded sequence and $\xi_0, \xi_1, \xi_2, \ldots$ is a martingale, then $\zeta_0, \zeta_1, \zeta_2, \ldots$ is a martingale (a fair game turns into a fair one no matter what you do);

2) If $\alpha_1, \alpha_2, \ldots$ is a non-negative bounded sequence and $\xi_0, \xi_1, \xi_2, \ldots$ is a supermartingale, then $\zeta_0, \zeta_1, \zeta_2, \ldots$ is a supermartingale (an unfavourable game turns into an unfavourable one).

3) If $\alpha_1, \alpha_2, \ldots$ is a non-negative bounded sequence and $\xi_0, \xi_1, \xi_2, \ldots$ is a submartingale, then $\zeta_0, \zeta_1, \zeta_2, \ldots$ is a submartingale (a favourable game turns into a favourable one).

Proof

Because α_n and ζ_{n-1} are $\mathcal{F}_{n-1}$-measurable, we can take them out of the expectation conditioned on $\mathcal{F}_{n-1}$ ('taking out what is known', Proposition 2.4). Thus, we obtain

$$\begin{aligned} E(\zeta_n|\mathcal{F}_{n-1}) &= E(\zeta_{n-1} + \alpha_n(\xi_n - \xi_{n-1})|\mathcal{F}_{n-1}) \\ &= \zeta_{n-1} + \alpha_n(E(\xi_n|\mathcal{F}_{n-1}) - \xi_{n-1}). \end{aligned}$$

If ξ_n is a martingale, then

$$\alpha_n(E(\xi_n|\mathcal{F}_{n-1}) - \xi_{n-1}) = 0,$$

which proves assertion 1). If ξ_n is a supermartingale and $\alpha_n \geq 0$, then

$$\alpha_n(E(\xi_n|\mathcal{F}_{n-1}) - \xi_{n-1}) \leq 0,$$

proving assertion 2). Finally, assertion 3) follows because

$$\alpha_n(E(\xi_n|\mathcal{F}_{n-1}) - \xi_{n-1}) \geq 0$$

if ξ_n is a submartingale and $\alpha_n \geq 0$. □

3.5 Stopping Times

In roulette and many other games of chance one usually has the option to quit at any time. The number of rounds played before quitting the game will be denoted by τ. It can be fixed, say, to be $\tau = 10$ if one decides in advance to stop playing after 10 rounds, no matter what happens. But in general the decision whether to quit or not will be made after each round depending on the knowledge accumulated so far. Therefore τ is assumed to be a random variable with values in the set $\{1, 2, \ldots\} \cup \{\infty\}$. Infinity is included to cover the theoretical possibility (and a dream scenario of some casinos) that the game never stops. At each step n one should be able to decide whether to stop playing or not, i.e. whether or not $\tau = n$. Therefore the event that $\tau = n$ should be in the σ-field $\mathcal{F}_n$ representing our knowledge at time n. This gives rise to the following definition.

Definition 3.7

A random variable τ with values in the set $\{1, 2, \ldots\} \cup \{\infty\}$ is called a *stopping time* (with respect to a filtration $\mathcal{F}_n$) if for each $n = 1, 2, \ldots$

$$\{\tau = n\} \in \mathcal{F}_n.$$

Exercise 3.8

Show that the following conditions are equivalent:

1) $\{\tau \leq n\} \in \mathcal{F}_n$ for each $n = 1, 2, \ldots$;

2) $\{\tau = n\} \in \mathcal{F}_n$ for each $n = 1, 2, \ldots$.

Hint Can you express $\{\tau \leq n\}$ in terms of the events $\{\tau = k\}$, where $k = 1, \ldots, n$? Can you express $\{\tau = n\}$ in terms of the events $\{\tau \leq k\}$, where $k = 1, \ldots, n$?

Example 3.5 (First hitting time)

Suppose that a coin is tossed repeatedly and you win or lose £1, depending on which way it lands. Suppose that you start the game with, say, £5 in your pocket and decide to play until you have £10 or you lose everything. If ξ_n is the amount you have at step n, then the time when you stop the game is

$$\tau = \min \{n : \xi_n = 10 \text{ or } 0\},$$

and is called the *first hitting time* (of 10 or 0 by the random sequence ξ_n). It is a stopping time in the sense of Definition 3.7 with respect to the filtration

$\mathcal{F}_n = \sigma(\xi_1, \ldots, \xi_n)$. This is because

$$\{\tau = n\} = \{0 < \xi_1 < 10\} \cap \cdots \cap \{0 < \xi_{n-1} < 10\} \cap \{\xi_n = 10 \text{ or } 0\}.$$

Now each of the sets on the right-hand side belongs to $\mathcal{F}_n$, so their intersection does too. This proves that

$$\{\tau = n\} \in \mathcal{F}_n$$

for each n, so τ is a stopping time.

Exercise 3.9

Let ξ_n be a sequence of random variables adapted to a filtration $\mathcal{F}_n$ and let $B \subset \mathbb{R}$ be a Borel set. Show that the *time of first entry* of ξ_n into B,

$$\tau = \min\{n : \xi_n \in B\}$$

is a stopping time.

Hint Example 3.5 covers the case when $B = (-\infty, 0] \cup [10, \infty)$. Extend the argument to an arbitrary Borel set B.

Let ξ_n be a sequence of random variables adapted to a filtration $\mathcal{F}_n$ and let τ be a stopping time (with respect to the same filtration). Suppose that ξ_n represents your winnings (or losses) after n rounds of a game. If you decide to quit after τ rounds, then your total winnings will be ξ_τ. In this case your winnings after n rounds will in fact be $\xi_{\tau\wedge n}$. Here $a \wedge b$ denotes the smaller of two numbers a and b,

$$a \wedge b = \min(a, b).$$

Definition 3.8

We call $\xi_{\tau\wedge n}$ the sequence *stopped* at τ. It is often denoted by ξ_n^τ. Thus, for each $\omega \in \Omega$

$$\xi_n^\tau(\omega) = \xi_{\tau(\omega)\wedge n}(\omega).$$

Exercise 3.10

Show that if ξ_n is a sequence of random variables adapted to a filtration $\mathcal{F}_n$, then so is the sequence $\xi_{\tau\wedge n}$.

Hint For any Borel set B express $\{\xi_{\tau\wedge n} \in B\}$ in terms of the events $\{\xi_k \in B\}$ and $\{\tau = k\}$, where $k = 1, \ldots, n$.

We already know that it is impossible to turn a fair game into an unfair one, an unfavourable game into a favourable one, or vice versa using a gambling strategy. The next proposition shows that this cannot be achieved using a stopping time either (essentially, because stopping is also a gambling strategy).

Proposition 3.2

Let τ be a stopping time.

1) If ξ_n is a martingale, then so is $\xi_{\tau\wedge n}$.

2) If ξ_n is a supermartingale, then so is $\xi_{\tau\wedge n}$.

3) If ξ_n is a submartingale, then so is $\xi_{\tau\wedge n}$.

Proof

This is in fact a consequence of Proposition 3.1. Given a stopping time τ, we put

$$\alpha_n = \begin{cases} 1 & \text{if } \tau \geq n, \\ 0 & \text{if } \tau < n. \end{cases}$$

We claim that α_n is a gambling strategy (that is, α_n is $\mathcal{F}_{n-1}$-measurable). This is because the inverse image $\{\alpha_n \in B\}$ of any Borel set $B \subset \mathbb{R}$ is equal to

$$\emptyset \in \mathcal{F}_{n-1}$$

if $0, 1 \notin B$, or to

$$\Omega \in \mathcal{F}_{n-1}$$

if $0, 1 \in B$, or to

$$\{\alpha_n = 1\} = \{\tau \geq n\} = \{\tau > n-1\} \in \mathcal{F}_{n-1}$$

if $1 \in B$ and $0 \notin B$, or to

$$\{\alpha_n = 0\} = \{\tau < n\} = \{\tau \leq n-1\} \in \mathcal{F}_{n-1}$$

if $1 \notin B$ and $0 \in B$. For this gambling strategy

$$\xi_{\tau\wedge n} = \alpha_1 (\xi_1 - \xi_0) + \cdots + \alpha_n (\xi_n - \xi_{n-1}).$$

Therefore Proposition 3.1 implies assertions 1), 2) and 3) above. □

Example 3.6

(You could try to beat the system if you had unlimited capital and unlimited time.) The following gambling strategy is called 'the martingale'. (Do not confuse this with the general definition of a martingale earlier in this section.) Suppose a coin is flipped repeatedly. Let us denote the outcomes by $\eta_1, \eta_2, \ldots$, which can take values $+1$ (heads) or -1 (tails). You wager £1 on heads. If you win, you quit. If you lose, you double the stake and play again. If you win this time round, you quit. Otherwise you double the stake once more, and so on. Thus, your gambling strategy is

$$\alpha_n = \begin{cases} 2^{n-1} & \text{if } \eta_1 = \cdots = \eta_{n-1} = \text{tails}, \\ 0 & \text{otherwise.} \end{cases}$$

Let us put

$$\zeta_n = \eta_1 + 2\eta_2 + \cdots + 2^{n-1}\eta_n$$

and consider the stopping time

$$\tau = \min\{n : \eta_n = \text{heads}\}.$$

Then $\zeta_{\tau \wedge n}$ will be your winnings after n rounds. It is a martingale (check it!).

It can be shown that $P\{\tau < \infty\} = 1$ (heads will eventually appear in the sequence $\eta_1, \eta_2, \ldots$ with probability one). Therefore it makes sense to consider ζ_τ. This would be your total winnings if you were able to continue to play the game no matter how long it takes for the first heads to appear. It would require unlimited time and capital. If you could afford these, you would be bound to win eventually because $\zeta_\tau = 1$ identically, since

$$-1 - 2 - \cdots - 2^{n-1} + 2^n = 1$$

for any n.

Exercise 3.11

Show that if a gambler plays 'the martingale', his expected loss just before the ultimate win is infinite, that is,

$$E(\zeta_{\tau-1}) = -\infty.$$

Hint What is the probability that the game will terminate at step n, i.e. that $\tau = n$? If $\tau = n$, what is $\zeta_{\tau-1}$ equal to? This will give you all possible values of $\zeta_{\tau-1}$ and their probabilities. Now compute the expectation of $\zeta_{\tau-1}$.

3.6 Optional Stopping Theorem

If ξ_n is a martingale, then, in particular,

$$E(\xi_n) = E(\xi_1)$$

for each n. Example 3.6 shows that $E(\xi_\tau)$ is not necessarily equal to $E(\xi_1)$ for a stopping time τ. However, if the equality

$$E(\xi_\tau) = E(\xi_1)$$

does hold, it can be very useful. The Optional Stopping Theorem provides sufficient conditions for this to happen.

Theorem 3.1 (Optional Stopping Theorem)

Let ξ_n be a martingale and τ a stopping time with respect to a filtration $\mathcal{F}_n$ such that the following conditions hold:

1) $\tau < \infty$ a.s.,

2) ξ_τ is integrable,

3) $E(\xi_n 1_{\{\tau>n\}}) \to 0$ as $n \to \infty$.

Then

$$E(\xi_\tau) = E(\xi_1).$$

Proof

Because

$$\xi_\tau = \xi_{\tau\wedge n} + (\xi_\tau - \xi_n)\, 1_{\{\tau>n\}},$$

it follows that

$$E(\xi_\tau) = E(\xi_{\tau\wedge n}) + E(\xi_\tau 1_{\{\tau>n\}}) - E(\xi_n 1_{\{\tau>n\}}).$$

Since $\xi_{\tau\wedge n}$ is a martingale by Proposition 3.2, the first term on the right-hand side is equal to

$$E(\xi_{\tau\wedge n}) = E(\xi_1).$$

The last term tends to zero by assumption 3). The middle term

$$E(\xi_\tau 1_{\{\tau>n\}}) = \sum_{k=n+1}^{\infty} E(\xi_k 1_{\{\tau=k\}})$$

tends to zero as $n \to \infty$ because the series

$$E(\xi_\tau) = \sum_{k=1}^{\infty} E(\xi_k 1_{\{\tau=k\}})$$

is convergent by 2). It follows that $E(\xi_\tau) = E(\xi_1)$, as required. □

Example 3.7 (Expectation of the first hitting time for a random walk)

Let ξ_n be a symmetric random walk as in Exercise 3.5 and let K be a positive integer. We define the first hitting time (of $\pm K$ by ξ_n) to be

$$\tau = \min\{n : |\xi_n| = K\}.$$

By Exercise 3.9 τ is a stopping time. By Exercise 3.5 we know that $\xi_n^2 - n$ is a martingale. If the Optional Stopping Theorem can be applied, then

$$E\left(\xi_\tau^2 - \tau\right) = E\left(\xi_1^2 - 1\right) = 0.$$

This allows us to find the expectation

$$E(\tau) = E(\xi_\tau^2) = K^2,$$

since $|\xi_\tau| = K$.

Let us verify conditions 1)–3) of the Optional Stopping Theorem.

1) We shall show that $P\{\tau = \infty\} = 0$. To this end we shall estimate $P\{\tau > 2Kn\}$. We can think of $2Kn$ tosses of a coin as n sequences of $2K$ tosses. A necessary condition for $\tau > 2Kn$ is that no one of these n sequences contains heads only. Therefore

$$P\{\tau > 2Kn\} \le \left(1 - \frac{1}{2^{2K}}\right)^n \to 0$$

as $n \to \infty$. Because $\{\tau > 2Kn\}$ for $n = 1, 2, \ldots$ is a contracting sequence of sets (i.e. $\{\tau > 2Kn\} \supset \{\tau > 2K(n+1)\}$), it follows that

$$\begin{aligned} P\{\tau = \infty\} &= \bigcap_{n=1}^{\infty} P\{\tau > 2Kn\} \\ &= \lim_{n\to\infty} P\{\tau > 2Kn\} = 0, \end{aligned}$$

completing the argument.

2) We need to show that

$$E\left(\left|\xi_\tau^2 - \tau\right|\right) < \infty.$$

Indeed,

$$\begin{aligned} E(\tau) &= \sum_{n=1}^{\infty} nP\{\tau = n\} \\ &= \sum_{n=0}^{\infty}\sum_{k=1}^{2K} (2Kn+k)\, P\{\tau = 2Kn+k\} \\ &\leq \sum_{n=0}^{\infty}\sum_{k=1}^{2K} 2K\,(n+1)\, P\{\tau > 2Kn\} \\ &\leq 4K^2 \sum_{n=0}^{\infty} (n+1)\left(1 - \frac{1}{2^{2K}}\right)^n \\ &< \infty, \end{aligned}$$

since the series $\sum_{n=1}^{\infty} (n+1)\, q^n$ is convergent for any $q \in (-1, 1)$. Here we have recycled the estimate for $P\{\tau > 2Kn\}$ used in 2). Moreover, $\xi_\tau^2 = K^2$, so

$$\begin{aligned} E\left(\left|\xi_\tau^2 - \tau\right|\right) &\leq E\left(\xi_\tau^2\right) + E\left(\tau\right) \\ &= K^2 + E(\tau) \\ &< \infty. \end{aligned}$$

3) Since $\xi_n^2 \leq K^2$ on $\{\tau > n\}$,

$$E\left(\xi_n^2 1_{\{\tau > n\}}\right) \leq K^2 P\{\tau > n\} \to 0$$

as $n \to \infty$. Moreover,

$$E\left(n 1_{\{\tau > n\}}\right) \leq E\left(\tau 1_{\{\tau > n\}}\right) \to 0$$

as $n \to \infty$. Convergence to 0 holds because $E(\tau) < \infty$ by 2) and $\{\tau > n\}$ is a contracting sequence of sets with intersection $\{\tau = \infty\}$ of measure zero. It follows that

$$E\left(\left(\xi_n^2 - n\right) 1_{\{\tau > n\}}\right) \to 0,$$

as required.

Exercise 3.12

Let ξ_n be a symmetric random walk and $\mathcal{F}_n$ the filtration defined in Exercise 3.5. Denote by τ the smallest n such that $|\xi_n| = K$ as in Example 3.7. Verify that

$$\zeta_n = (-1)^n \cos\left[\pi\left(\xi_n + K\right)\right]$$

is a martingale (see Exercise 3.6). Then show that ζ_n and τ satisfy the conditions of the Optional Stopping Theorem and apply the theorem to find $E[(-1)^\tau]$.

Hint The equality $\zeta_\tau = (-1)^\tau$ is a key to computing $E[(-1)^\tau]$ with the aid of the Optimal Stopping Theorem. The first two conditions of this theorem are either obvious in the case in hand or have been verified elsewhere in this chapter. To make sure that condition 3) holds it may be helpful to show that

$$\left|E(\zeta_n 1_{\{\tau>n\}})\right| \leq P\{\tau > n\}.$$

Use Jensen's inequality with convex function $\varphi(x) = |x|$ to estimate the left-hand side. Do not forget to verify that ζ_n is a martingale in the first place.

3.7 Solutions

Solution 3.1

A belongs to $\mathcal{F}_{11}$, but not to $\mathcal{F}_{10}$. The smallest n is 11.
B does not belong to $\mathcal{F}_n$ for any n. There is no smallest n such that $B \in \mathcal{F}_n$.
C belongs to $\mathcal{F}_{100}$, but not to $\mathcal{F}_{99}$. The smallest n is 100.
Since $D = \emptyset$, it belongs to $\mathcal{F}_n$ for each $n = 1, 2, \ldots$. Here the smallest n is 1.

Solution 3.2

Because the sequence of random variables $\xi_1, \xi_2, \ldots$ is adapted to the filtration $\mathcal{G}_1, \mathcal{G}_2, \ldots$, it follows that ξ_n is $\mathcal{G}_n$-measurable for each n. But

$$\mathcal{G}_1 \subset \mathcal{G}_2 \subset \cdots,$$

so $\xi_1, \ldots, \xi_n$ are $\mathcal{G}_n$-measurable for each n. As a consequence,

$$\mathcal{F}_n = \sigma(\xi_1, \ldots, \xi_n) \subset \mathcal{G}_n$$

for each n.

Solution 3.3

Taking the expectation on both sides of the equality

$$\xi_n = E(\xi_{n+1}|\mathcal{F}_n),$$

we obtain

$$E(\xi_n) = E(E(\xi_{n+1}|\mathcal{F}_n)) = E(\xi_{n+1})$$

for each n. This proves the claim.

Solution 3.4

The random variables ξ_n are integrable because ξ_n is a martingale with respect to $\mathcal{F}_n$. Since $\mathcal{G}_n$ is the σ-field generated by $\xi_1, \ldots, \xi_n$, it follows that ξ_n is adapted to $\mathcal{G}_n$. Finally, since $\mathcal{G}_n \subset \mathcal{F}_n$,

$$\begin{aligned} \xi_n &= E(\xi_n|\mathcal{G}_n) \\ &= E(E(\xi_{n+1}|\mathcal{F}_n)|\mathcal{G}_n) \\ &= E(\xi_{n+1}|\mathcal{G}_n) \end{aligned}$$

by the tower property of conditional expectation (Proposition 2.4). This proves that ξ_n is a martingale with respect to $\mathcal{G}_n$.

Solution 3.5

Because

$$\xi_n^2 - n = (\eta_1 + \cdots + \eta_n)^2 - n$$

is a function of $\eta_1, \ldots, \eta_n$, it is measurable with respect to the σ-field $\mathcal{F}_n$ generated by $\eta_1, \ldots, \eta_n$, i.e. $\xi_n^2 - n$ is adapted to $\mathcal{F}_n$. Since

$$|\xi_n| = |\eta_1 + \cdots + \eta_n| \leq |\eta_1| + \cdots |\eta_n| = n,$$

it follows that

$$E(|\xi_n^2 - n|) \leq E(\xi_n^2) + n \leq n^2 + n < \infty,$$

so $\xi_n^2 - n$ is integrable for each n. Because

$$\xi_{n+1}^2 = \eta_{n+1}^2 + 2\eta_{n+1}\xi_n + \xi_n^2,$$

where ξ_n and ξ_n^2 are $\mathcal{F}_n$-measurable and η_{n+1} is independent of $\mathcal{F}_n$, we can use Proposition 2.4 ('taking out what is known' and 'independent condition drops out') to obtain

$$\begin{aligned} E(\xi_{n+1}^2|\mathcal{F}_n) &= E(\eta_{n+1}^2|\mathcal{F}_n) + 2E(\eta_{n+1}\xi_n|\mathcal{F}_n) + E(\xi_n^2|\mathcal{F}_n) \\ &= E(\eta_{n+1}^2) + 2\xi_n E(\eta_{n+1}) + \xi_n^2 \\ &= 1 + \xi_n^2. \end{aligned}$$

This implies that

$$E(\xi_{n+1}^2 - n - 1|\mathcal{F}_n) = \xi_n^2 - n,$$

so $\xi_n^2 - n$ is a martingale.

Solution 3.6

Being a function of ξ_n, the random variable ζ_n is $\mathcal{F}_n$-measurable for each n, since ξ_n is. Because $|\zeta_n| \leq 1$, it is clear that ζ_n is integrable. Because η_{n+1} is independent of $\mathcal{F}_n$ and ξ_n is $\mathcal{F}_n$-measurable, it follows that

$$\begin{aligned} E(\zeta_{n+1}|\mathcal{F}_n) &= E\left((-1)^{n+1}\cos[\pi(\xi_n+\eta_{n+1})]|\mathcal{F}_n\right) \\ &= (-1)^{n+1}E\left(\cos\left(\pi\xi_n\right)\cos\left(\pi\eta_{n+1}\right)|\mathcal{F}_n\right) \\ &\quad -(-1)^{n+1}E\left(\sin\left(\pi\xi_n\right)\sin\left(\pi\eta_{n+1}\right)|\mathcal{F}_n\right) \\ &= (-1)^{n+1}\cos\left(\pi\xi_n\right)E\left(\cos\left(\pi\eta_{n+1}\right)\right) \\ &\quad -(-1)^{n+1}\sin\left(\pi\xi_n\right)E\left(\sin\left(\pi\eta_{n+1}\right)\right) \\ &= (-1)^n\cos\left(\pi\xi_n\right) \\ &= \zeta_n, \end{aligned}$$

using the formula

$$\cos(\alpha+\beta) = \cos\alpha\cos\beta - \sin\alpha\sin\beta.$$

To compute $E\left(\cos\left(\pi\eta_{n+1}\right)\right)$ and $E\left(\sin\left(\pi\eta_{n+1}\right)\right)$ observe that $\eta_{n+1} = 1$ or -1 and

$$\begin{aligned} \cos\pi &= \cos(-\pi) = -1, \\ \sin\pi &= \sin(-\pi) = 0. \end{aligned}$$

It follows that ζ_n is a martingale with respect to the filtration $\mathcal{F}_n$.

Solution 3.7

If ξ_n is adapted to $\mathcal{F}_n$, then so is ξ_n^2. Since $\xi_n = E\left(\xi_{n+1}|\mathcal{F}_n\right)$ for each n and $\varphi(x) = x^2$ is a convex function, we can apply Jensen's inequality (Theorem 2.2) to obtain

$$\xi_n^2 = \left[E\left(\xi_{n+1}|\mathcal{F}_n\right)\right]^2 \leq E\left(\xi_{n+1}^2|\mathcal{F}_n\right)$$

for each n. This means that ξ_n^2 is a submartingale with respect to $\mathcal{F}_n$.

Solution 3.8

1)$\Rightarrow$2). If τ has property 1), then

$$\{\tau \leq n\} \in \mathcal{F}_n$$

and

$$\{\tau \leq n-1\} \in \mathcal{F}_{n-1} \subset \mathcal{F}_n,$$

so

$$\{\tau = n\} = \{\tau \leq n\} \setminus \{\tau \leq n-1\} \in \mathcal{F}_n.$$

2)$\Rightarrow$1). If τ has property 2), then

$$\{\tau = k\} \in \mathcal{F}_k \subset \mathcal{F}_n$$

for each $k = 1, \ldots, n$. Therefore

$$\{\tau \leq n\} = \{\tau = 1\} \cup \cdots \cup \{\tau = n\} \in \mathcal{F}_n.$$

Solution 3.9

If

$$\tau = \min\{n : \xi_n \in B\},$$

then for any n

$$\{\tau = n\} = \{\xi_1 \notin B\} \cap \cdots \cap \{\xi_{n-1} \notin B\} \cap \{\xi_n \in B\}.$$

Because B is a Borel set, each of the sets on the right-hand side belongs to the σ-field $\mathcal{F}_n = \sigma(\xi_1, \ldots, \xi_n)$, and their intersection does too. This proves that $\{\tau = n\} \in \mathcal{F}_n$ for each n, so τ is a stopping time.

Solution 3.10

Let $B \subset \mathbb{R}$ be a Borel set. We can write

$$\{\xi_{\tau\wedge n} \in B\} = \{\xi_n \in B, \tau > n\} \cup \bigcup_{k=1}^{n} \{\xi_k \in B, \tau = k\},$$

where

$$\{\xi_n \in B, \tau > n\} = \{\xi_n \in B\} \cap \{\tau > n\} \in \mathcal{F}_n$$

and for each $k = 1, \ldots, n$

$$\{\xi_k \in B, \tau = k\} = \{\xi_k \in B\} \cap \{\tau = k\} \in \mathcal{F}_k \subset \mathcal{F}_n.$$

It follows that for each n

$$\{\xi_{\tau\wedge n} \in B\} \in \mathcal{F}_n,$$

as required.

Solution 3.11

The probability that 'the martingale' terminates at step n is

$$P\{\tau = n\} = \frac{1}{2^n}$$

($n - 1$ tails followed by heads at step n). Therefore

$$\begin{aligned} E(\zeta_{\tau-1}) &= \sum_{n=1}^{\infty} \zeta_{n-1} P\{\tau = n\} \\ &= \sum_{n=1}^{\infty} \left(-1 - 2 - \cdots - 2^{n-2}\right) \frac{1}{2^n} \\ &= -\sum_{n=1}^{\infty} \frac{2^{n-1} - 1}{2^n} = -\infty. \end{aligned}$$

Solution 3.12

The proof that ζ_n is a martingale is almost the same as in Exercise 3.6. We need to verify that ζ_n and τ satisfy conditions 1)–3) of the Optional Stopping Theorem.

Condition 1) has in fact been verified in Example 3.7.

Condition 2) holds because $|\zeta_\tau| \leq 1$, so $E(|\zeta_\tau|) \leq 1 < \infty$.

To verify condition 3) observe that $|\zeta_n| \leq 1$ for all n, so

$$\begin{aligned} \left|E(\zeta_n 1_{\{\tau>n\}})\right| &\leq E(|\zeta_n| 1_{\{\tau>n\}}) \\ &\leq E(1_{\{\tau>n\}}) \\ &= P\{\tau > n\}. \end{aligned}$$

The family of events $\{\tau > n\}$, $n = 1, 2, \ldots$ is a contracting one with intersection $\{\tau = \infty\}$. It follows that

$$\left|E(\zeta_n 1_{\{\tau>n\}})\right| \leq P\{\tau > n\} \searrow P\{\tau = \infty\}$$

as $n \to \infty$. But

$$P\{\tau = \infty\} = 0$$

by 1), completing the proof.

The Optional Stopping Theorem implies that

$$E(\zeta_\tau) = E(\zeta_1)$$

Because $\xi_\tau = K$ or $-K$, we have

$$\zeta_\tau = (-1)^\tau \cos[\pi(K + \xi_\tau)] = (-1)^\tau.$$

Let us compute

$$\begin{aligned} E(\zeta_1) &= -\frac{1}{2}\left(\cos\left[\pi(1+K)\right] + \cos\left[\pi(-1+K)\right]\right) \\ &= \cos(\pi K) = (-1)^K. \end{aligned}$$

It follows that

$$E[(-1)^\tau] = (-1)^K.$$

4
Martingale Inequalities and Convergence

Results on the convergence of martingales provide an insight into their structure and have a multitude of applications. They also provide an important interpretation of martingales. Namely, it turns out that a large class of martingales can be represented in the form

$$\xi_n = E(\xi|\mathcal{F}_n), \tag{4.1}$$

where $\xi = \lim_n \xi_n$ is an integrable random variable and $\mathcal{F}_1, \mathcal{F}_2, \ldots$ is the filtration generated by $\xi_1, \xi_2, \ldots$, see Theorem 4.4 below. This makes it possible to think of $\xi_1, \xi_2, \ldots$ as the results of a series of imperfect observations of some random quantity ξ. As n increases, the accumulated knowledge $\mathcal{F}_n$ about ξ increases and ξ_n becomes a better approximation, approaching the observed quantity ξ in the limit.

We shall begin with a few classical inequalities for martingales, known as the Doob inequalities. They provide the tools we shall need to study the convergence of martingales and, later on, the properties of stochastic integrals. Then we shall present a classical result known as Doob's Martingale Convergence Theorem, which provides the limit $\lim_n \xi_n$ of a martingale. However, Doob's theorem has one inconvenient feature. It guarantees only that ξ_n converges a.s., even though the limit is known to be an integrable random variable. However, to obtain (4.1) we need convergence in L^1, which gives rise to a condition called uniform integrability. This condition and its consequences, including (4.3), will be studied in the second section. Finally, as an example of an application reaching beyond the theory of martingales, we present an elegant proof of Kolmogorov's 0–1 law.

4.1 Doob's Martingale Inequalities

Proposition 4.1 (Doob's maximal inequality)

Suppose that $\xi_n, n \in \mathbb{N}$, is a non-negative submartingale (with respect to a filtration $\mathcal{F}_n$). Then for any $\lambda > 0$

$$\lambda P\left(\max_{k\leq n} \xi_k \geq \lambda\right) \leq E\left(\xi_n 1_{\{\max_{k\leq n} \xi_k \geq \lambda\}}\right),$$

where 1_A is the characteristic function of a set A.

Proof

We put $\xi_n^* = \max_{k\leq n} \xi_k$ for brevity. For $\lambda > 0$ let us define

$$\tau = \min\{k \leq n : \xi_k \geq \lambda\},$$

if there is a $k \leq n$ such that $\xi_k \geq \lambda$, and $\tau = n$ otherwise. Then τ is a stopping time such that $\tau \leq n$ a.s. Since ξ_n is a submartingale,

$$E(\xi_n) \geq E(\xi_\tau).$$

But

$$E(\xi_\tau) = E\left(\xi_\tau 1_{\{\xi_n^* \geq \lambda\}}\right) + E\left(\xi_\tau 1_{\{\xi_n^* < \lambda\}}\right).$$

Observe that if $\xi_n^* \geq \lambda$, then $\xi_\tau \geq \lambda$. Moreover, if $\xi_n^* < \lambda$, then $\tau = n$, and so $\xi_\tau = \xi_n$. Therefore

$$E(\xi_n) \geq E(\xi_\tau) \geq \lambda P(\xi_n^* \geq \lambda) + E\left(\xi_n 1_{\{\xi_n^* < \lambda\}}\right),$$

It follows that

$$\lambda P(\xi_n^* \geq \lambda) \leq E(\xi_n) - E\left(\xi_n 1_{\{\xi_n^* < \lambda\}}\right) = E\left(\xi_n 1_{\{\xi_n^* \geq \lambda\}}\right),$$

completing the proof. □

Theorem 4.1 (Doob's maximal L^2 inequality)

If $\xi_n, n \in \mathbb{N}$, is a non-negative square integrable submartingale (with respect to a filtration $\mathcal{F}_n$), then

$$E\left|\max_{k\leq n} \xi_k\right|^2 \leq 4E|\xi_n|^2. \tag{4.2}$$

Proof

Put $\xi_n^* = \max_{k\leq n} \xi_k$. By Exercise 1.9, Proposition 4.1, the Fubini theorem and finally the Cauchy–Schwarz inequality

$$\begin{aligned} E\left|\xi_n^*\right|^2 &= 2\int_0^\infty tP\left(\xi_n^* > t\right) dt \leq 2\int_0^\infty E\left(\xi_n 1_{\{\xi_n^* \geq t\}}\right) dt \\ &= 2\int_0^\infty \left(\int_{\{\xi_n^* \geq t\}} \xi_n \, dP\right) dt = 2\int_\Omega \xi_n \left(\int_0^{\xi_n^*} dt\right) dP \\ &= 2\int_\Omega \xi_n \xi_n^* dP = 2E\left(\xi_n \xi_n^*\right) \leq 2\left(E\left|\xi_n\right|^2\right)^{1/2} \left(E\left|\xi_n^*\right|^2\right)^{1/2}. \end{aligned}$$

Dividing by $\left(E\left|\xi_n^*\right|^2\right)^{1/2}$, we get (4.2). □

The proof of Doob's Convergence Theorem in the next section hinges on an inequality involving the number of *upcrossings.*

Definition 4.1

Given an adapted sequence of random variables $\xi_1, \xi_2, \ldots$ and two real numbers $a < b$, we define a gambling strategy $\alpha_1, \alpha_2, \ldots$ by putting

$$\alpha_1 = 0$$

and for $n = 1, 2, \ldots$

$$\alpha_{n+1} = \begin{cases} 1 & \text{if } \alpha_n = 0 \text{ and } \xi_n < a, \\ 1 & \text{if } \alpha_n = 1 \text{ and } \xi_n \leq b, \\ 0 & \text{otherwise.} \end{cases}$$

It will be called the *upcrossings strategy.* Each $k = 1, 2, \ldots$ such that $\alpha_k = 1$ and $\alpha_{k+1} = 0$ will be called an *upcrossing* of the interval $[a, b]$. The upcrossings form a (finite or infinite) increasing sequence

$$u_1 < u_2 < \cdots .$$

The number of upcrossings made up to time n, that is, the largest k such that $u_k \leq n$ will be denoted by $U_n[a, b]$ (we put $U_n[a, b] = 0$ if no such k exists).

The meaning of the above definition is this. Initially, we refrain from playing the game and wait until ξ_n becomes less than a. As soon as this happens, we start playing unit stakes at each round of the game and continue until ξ_n becomes greater than b. At this stage we refrain from playing again, wait until ξ_n becomes less than a, and so on. The strategy α_n is defined in such a way that $\alpha_n = 0$ whenever we refrain from playing the nth game, and $\alpha_n = 1$ otherwise. During each run of consecutive games with $\alpha_n = 1$ the process ξ_n crosses the interval $[a, b]$, starting below a and finishing above b. This is what is meant by an upcrossing. Observe that each upcrossing will increase our total winnings by at least $b - a$. For convenience, we identify each upcrossing with its last step k, such that $\alpha_k = 1$ and $\alpha_{k+1} = 0$. A typical sample path of the upcrossings strategy is shown in Figure 4.1.

Exercise 4.1

Verify that the upcrossings strategy α_n is indeed a gambling strategy.

Hint You want to prove that α_n is $\mathcal{F}_{n-1}$-measurable for each n. Since the upcrossings strategy is defined by induction, a proof by induction on n may be your best bet.

Lemma 4.1 (Upcrossings Inequality)

If $\xi_1, \xi_2, \ldots$ is a supermartingale and $a < b$, then

$$(b - a)E(U_n[a, b]) \leq E((\xi_n - a)^-).$$

By x^- we denote the negative part of a real number x, i.e. $x^- = \max\{0, -x\}$.

Proof

Let

$$\zeta_n = \alpha_1 (\xi_1 - \xi_0) + \cdots + \alpha_n (\xi_n - \xi_{n-1})$$

be the total winnings at step $n = 1, 2, \ldots$ if the upcrossings strategy is followed, see Figure 4.1. It will be convenient to put $\zeta_0 = 0$. By Proposition 3.1 (one cannot beat the system using a gambling strategy) ζ_n is a supermartingale.

Let us fix an n and put $k = U_n[a, b]$, so that

$$0 < u_1 < u_2 < \cdots < u_k \leq n$$

Clearly, each upcrossing increases the total winnings by $b - a$,

$$\zeta_{u_i} - \zeta_{u_{i-1}} \geq b - a$$

for $i = 1, \ldots, k$. (We put $u_0 = 0$ for simplicity.) Moreover,

$$\zeta_n - \zeta_{u_k} \geq -(\xi_n - a)^- .$$

It follows that

$$\zeta_n \geq (b - a)U_n[a, b] - (\xi_n - a)^- .$$

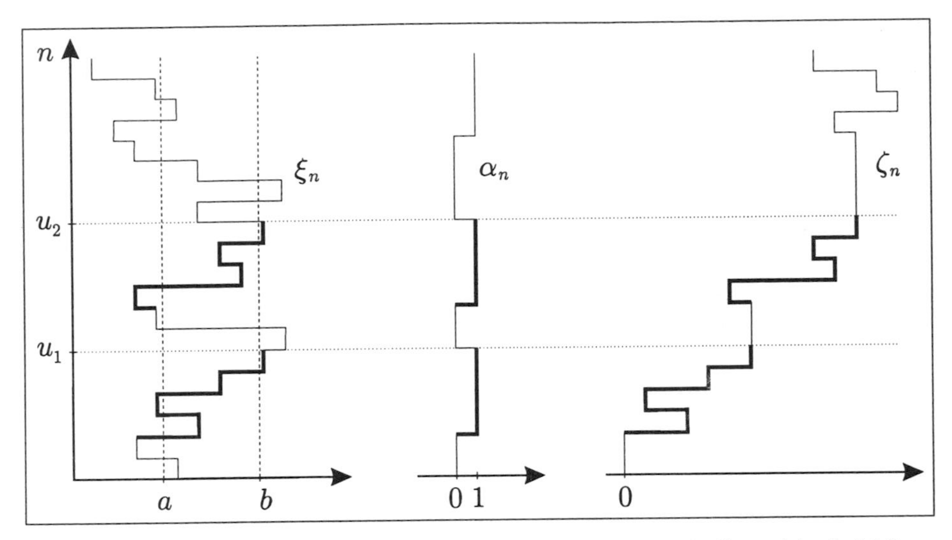

Figure 4.1. Typical paths of ξ_n, α_n and ζ_n; upcrossings are indicated by bold lines

taking the expectation on both sides, we get

$$E(\zeta_n) \geq (b-a)E(U_n[a,b]) - E((\xi_n - a)^-).$$

But ζ_n is a supermartingale, so

$$0 = E(\zeta_1) \geq E(\zeta_n),$$

which proves the Upcrossings Inequality. □

4.2 Doob's Martingale Convergence Theorem

Theorem 4.2 (Doob's Martingale Convergence Theorem)

Suppose that $\xi_1, \xi_2, \ldots$ is a supermartingale (with respect to a filtration $\mathcal{F}_1, \mathcal{F}_2, \ldots$) such that

$$\sup_n E(|\xi_n|) < \infty.$$

Then there is an integrable random variable ξ such that

$$\lim_{n\to\infty} \xi_n = \xi \quad \text{a.s.}$$

Remark 4.1

In particular, the theorem is valid for martingales because every martingale is a supermartingale. It is also valid for submartingales, since ξ_n is a submartingale if and only if $-\xi_n$ is a supermartingale.

Remark 4.2

Observe that even though all the ξ_n as well as the limit ξ are integrable random variables, it is claimed only that ξ_n trends to ξ a.s. Note that no convergence in L^1 is asserted.

Proof (of Doob's Martingale Convergence Theorem)

By the Upcrossings Inequality

$$E\left(U_n[a,b]\right) \leq \frac{E\left(\left(\xi_n - a\right)^-\right)}{b-a} \leq \frac{M+|a|}{b-a} < \infty,$$

where

$$M = \sup_n E\left(|\xi_n|\right) < \infty.$$

Since $U_n[a,b]$ is a non-decreasing sequence, it follows that

$$E\left(\lim_{n\to\infty} U_n[a,b]\right) = \lim_{n\to\infty} E\left(U_n[a,b]\right) \leq \frac{M+|a|}{b-a} < \infty.$$

This implies that

$$P\left\{\lim_{n\to\infty} U_n[a,b] < \infty\right\} = 1.$$

for any $a < b$. Because the set of rational numbers is countable, it follows that

$$P\left\{\lim_{n\to\infty} U_n[a,b] < \infty \text{ for all rational } a < b\right\} = 1. \tag{4.3}$$

(The intersection of countably many events has probability one if each of these events has probability one.)

We claim that the sequence ξ_n converges a.s. to a limit ξ. Suppose that

$$\liminf_n \xi_n < \limsup_n \xi_n.$$

Then $U_n[a,b] \nearrow \infty$ if

$$\liminf_n \xi_n < a < b < \limsup_n \xi_n.$$

Because a, b can be chosen to be rational numbers, this contradicts (4.3), proving the claim.

It remains to show that the limit ξ is an integrable random variable. By Fatou's lemma

$$\begin{aligned} E(|\xi|) &= E\left(\liminf_n |\xi_n|\right) \\ &\leq \liminf_n E(|\xi_n|) \\ &< \sup_n E(|\xi_n|) < \infty. \end{aligned}$$

This completes the proof. □

Exercise 4.2

Show that if ξ_n is a non-negative supermartingale, then it converges a.s. to an integrable random variable.

Hint To apply Doob's Theorem all you need to verify is that the sequence ξ_n is bounded in L^1, i.e. the supremum of $E(|\xi_n|)$ is less than ∞.

4.3 Uniform Integrability and L^1 Convergence of Martingales

The conditions of Doob's theorem imply pointwise (a.s.) convergence of martingales. In this section we shall study convergence in L^1. To this end we introduce a stronger condition called *uniform integrability*. Proposition 4.2 shows that it is a necessary condition for L^1 convergence. In Theorem 4.2 we prove that uniform integrability is in fact sufficient for a martingale to converge in L^1. This enables us to show that each integrable martingale is of the form $E(\xi|\mathcal{F}_n)$. As an application we give a martingale proof of Kolmogorov's 0–1 law.

Exercise 4.3

Show that a random variable ξ is integrable if and only if for every $\varepsilon > 0$ there exists an $M > 0$ such that

$$\int_{\{|\xi|>M\}} |\xi| \, dP < \varepsilon.$$

Hint Split Ω into two sets: $\{|\xi| > M\}$ and $\{|\xi| \leq M\}$. The integrals of $|\xi|$ over these sets must add up to $E(|\xi|)$. As M increases, one of the integrals increases, while the other one decreases. Investigate their limits as $M \to \infty$.

Thus, for any sequence ξ_n of integrable random variables and any $\varepsilon > 0$ there is a sequence of numbers $M_n > 0$ such that

$$\int_{\{|\xi_n|>M_n\}} |\xi_n| \, dP < \varepsilon.$$

If the M_n are independent of n, then we say that the sequence ξ_n is uniformly integrable.

Definition 4.2

A sequence $\xi_1, \xi_2, \ldots$ of random variables is called *uniformly integrable* if for every $\varepsilon > 0$ there exists an $M > 0$ such that

$$\int_{\{|\xi_n|>M\}} |\xi_n| \, dP < \varepsilon$$

for all $n = 1, 2, \ldots$.

Exercise 4.4

Let $\Omega = [0,1]$ with the σ-field of Borel sets and Lebesgue measure. Take

$$\xi_n = n 1_{(0,\frac{1}{n})}.$$

Show that the sequence $\xi_1, \xi_2, \ldots$ is not uniformly integrable.

Hint What is the integral of ξ_n over $\{\xi_n > M\}$ if $n > M$?

Proposition 4.2

Uniform integrability is a necessary condition for a sequence $\xi_1, \xi_2, \ldots$ of integrable random variables to converge in L^1.

Lemma 4.2

If ξ is integrable, then for every $\varepsilon > 0$ there is a $\delta > 0$ such that

$$P(A) < \delta \Longrightarrow \int_A |\xi| \, dP < \varepsilon.$$

Proof (of Lemma 4.2)

Let $\varepsilon > 0$. Since ξ is integrable, by Exercise 4.3 there is an $M > 0$ such that

$$\int_{\{|\xi|>M\}} |\xi| \, dP < \frac{\varepsilon}{2}.$$

Now

$$\begin{aligned}\int_A |\xi|\, dP &= \int_{A\cap\{|\xi|\le M\}} |\xi|\, dP + \int_{A\cap\{|\xi|>M\}} |\xi|\, dP \\ &\le \int_A M\, dP + \int_{\{|\xi|>M\}} |\xi|\, dP \\ &< MP(A) + \frac{\varepsilon}{2}.\end{aligned}$$

Let $\delta = \frac{\varepsilon}{2M}$. Then

$$P(A) < \delta \Longrightarrow \int_A |\xi|\, dP < \varepsilon,$$

as required. □

Exercise 4.5

Let ξ be an integrable random variable and $\mathcal{F}_1, \mathcal{F}_2, \ldots$ a filtration. Show that $E(\xi|\mathcal{F}_n)$ is a uniformly integrable martingale.

Hint Use Lemma 4.2.

Proof (of Proposition 4.2)

Suppose that $\xi_n \to \xi$ in L^1, i.e. $E|\xi_n - \xi| \to 0$. We take any $\varepsilon > 0$. There is an integer N such that

$$n \ge N \Longrightarrow E|\xi_n - \xi| < \frac{\varepsilon}{2}.$$

By Lemma 4.2 there is a $\delta > 0$ such that

$$P(A) < \delta \Longrightarrow \int_A |\xi|\, dP < \frac{\varepsilon}{2}.$$

Taking a smaller $\delta > 0$ if necessary, we also have

$$P(A) < \delta \Longrightarrow \int_A |\xi_n|\, dP < \varepsilon \quad \text{for } n = 1, \ldots, N.$$

We claim that there is an $M > 0$ such that

$$P\{|\xi_n| > M\} < \delta$$

for all n. Indeed, since

$$E(|\xi_n|) \ge \int_{\{|\xi_n|>M\}} |\xi_n|\, dP \ge MP\{|\xi_n| > M\},$$

it suffices to take

$$M = \frac{1}{\delta} \sup_n E(|\xi_n|).$$

(Because the sequence ξ_n converges in L^1, it is bounded in L^1, so the supremum is $< \infty$.)

Now, since $P\{|\xi_n| > M\} < \delta$,

$$\begin{aligned}\int_{\{|\xi_n|>M\}} |\xi_n|\, dP &\leq \int_{\{|\xi_n|>M\}} |\xi|\, dP + \int_{\{|\xi_n|>M\}} |\xi_n - \xi|\, dP \\ &\leq \int_{\{|\xi_n|>M\}} |\xi|\, dP + E\left(|\xi_n - \xi|\right) \\ &< \frac{\varepsilon}{2} + \frac{\varepsilon}{2} = \varepsilon.\end{aligned}$$

for any $n > N$ and

$$\int_{\{|\xi_n|>M\}} |\xi_n|\, dP < \varepsilon$$

for any $n = 1, \ldots, N$, completing the proof. □

Exercise 4.6

Show that a uniformly integrable sequence of random variables is bounded in L^1, i.e.

$$\sup_n E\left(|\xi_n|\right) < \infty.$$

Hint Write $E\left(|\xi_n|\right)$ as the sum of the integrals of $|\xi_n|$ over $\{|\xi_n| > M\}$ and $\{|\xi_n| \leq M\}$.

Exercise 4.6 implies that each uniformly integrable martingale satisfies the conditions of Doob's theorem. Therefore it converges a.s. to an integrable random variable. We shall show that in fact it converges in L^1.

Theorem 4.3

Every uniformly integrable supermartingale (submartingale) ξ_n converges in L^1.

Proof

By Exercise 4.6 the sequence ξ_n is bounded in L^1, so it satisfies the conditions of Theorem 4.2 (Doob's Martingale Convergence Theorem). Therefore, there is an integrable random variable ξ such that $\xi_n \to \xi$ a.s. We can assume without loss of generality that $\xi = 0$ (since $\xi_n - \xi$ can be taken in place of ξ_n). That is to say,

$$P\left\{\lim_n \xi_n = 0\right\} = 1.$$

It follows that $\xi_n \to 0$ in probability, i.e. for any $\varepsilon > 0$

$$P\{|\xi_n| > \varepsilon\} \to 0$$

as $n \to \infty$. This is because by Fatou's lemma

$$\begin{aligned}\limsup_n P\{|\xi_n| > \varepsilon\} &\le P\left(\limsup_n \{|\xi_n| > \varepsilon\}\right) \\ &\le P\left(\Omega \setminus \left\{\lim_n \xi_n = 0\right\}\right) \\ &= 0.\end{aligned}$$

Let $\varepsilon > 0$. By uniform integrability there is an $M > 0$ such that

$$\int_{\{|\xi_n| > M\}} |\xi_n| \, dP \le \frac{\varepsilon}{3}$$

for all n. Since $\xi_n \to 0$ in probability, there is an integer N such that if $n > N$, then

$$P\left\{|\xi_n| > \frac{\varepsilon}{3}\right\} < \frac{\varepsilon}{3M}.$$

We can assume without loss of generality that $M > \frac{\varepsilon}{3}$. Then

$$\begin{aligned}E(|\xi_n|) &= \int_{\{|\xi_n| > M\}} |\xi_n| \, dP + \int_{\{M \ge |\xi_n| > \frac{\varepsilon}{3}\}} |\xi_n| \, dP \\ &\quad + \int_{\{\frac{\varepsilon}{3} \ge |\xi_n|\}} |\xi_n| \, dP \\ &\le \frac{\varepsilon}{3} + MP\left\{|\xi_n| > \frac{\varepsilon}{3}\right\} + \frac{\varepsilon}{3} P\left\{\frac{\varepsilon}{3} \ge |\xi_n|\right\} \\ &< \varepsilon\end{aligned}$$

for all $n > N$. This proves that $E(|\xi_n|) \to 0$, that is, $\xi_n \to 0$ in L^1. □

Theorem 4.4

Let ξ_n be a uniformly integrable martingale. Then

$$\xi_n = E(\xi | \mathcal{F}_n),$$

where $\xi = \lim_n \xi_n$ is the limit of ξ_n in L^1 and $\mathcal{F}_n = \sigma(\xi_1, \ldots, \xi_n)$ is the filtration generated by ξ_n.

Proof

For any $m > n$

$$E(\xi_m | \mathcal{F}_n) = \xi_n,$$

i.e. for any $A \in \mathcal{F}_n$

$$\int_A \xi_m \, dP = \int_A \xi_n \, dP.$$

Let n be an arbitrary integer and let $A \in \mathcal{F}_n$. For any $m > n$

$$\begin{aligned}\left|\int_A (\xi_n - \xi)\, dP\right| &= \left|\int_A (\xi_m - \xi)\, dP\right| \\ &\leq \int_A |\xi_m - \xi|\, dP \\ &\leq E(|\xi_m - \xi|) \to 0\end{aligned}$$

as $m \to \infty$. It follows that

$$\int_A \xi_n\, dP = \int_A \xi\, dP$$

for any $A \in \mathcal{F}_n$, so $\xi_n = E(\xi|\mathcal{F}_n)$. □

Exercise 4.7

Show that if ξ_n is a martingale and $\xi_n \to a$ in L^1 for some $a \in \mathbb{R}$, then $\xi_n = a$ a.s. for each n.

Hint Apply Theorem 4.4.

Theorem 4.5 (Kolmogorov's 0–1 Law)

Let $\eta_1, \eta_2, \ldots$ be a sequence of independent random variables. We define the *tail* σ*-field*

$$\mathcal{T} = \mathcal{T}_1 \cap \mathcal{T}_2 \cap \ldots,$$

where $\mathcal{T}_n = \sigma(\eta_n, \eta_{n+1}, \ldots)$. Then

$$P(A) = 0 \text{ or } 1$$

for any $A \in \mathcal{T}$.

Proof

Take any $A \in \mathcal{T}$ and define

$$\xi_n = E(1_A|\mathcal{F}_n),$$

where $\mathcal{F}_n = \sigma(\eta_1, \ldots, \eta_n)$. By Exercise 4.5 ξ_n is a uniformly integrable martingale, so $\xi_n \to \xi$ in L^1. By Theorem 4.4

$$E(\xi|\mathcal{F}_n) = E(1_A|\mathcal{F}_n)$$

for all n. Both $\xi = \lim_n \xi_n$ and 1_A are measurable with respect to the σ-field

$$\mathcal{F}_\infty = \sigma(\eta_1, \eta_2, \ldots).$$

Because this σ-field is generated by the family of sets $\mathcal{F}_1 \cup \mathcal{F}_2 \cup \cdots$ it follows that $\xi = 1_A$ a.s.

Since η_n is a sequence of independent random variables, the σ-fields $\mathcal{F}_n$ and $\mathcal{T}_{n+1}$ are independent. Because $\mathcal{T} \subset \mathcal{T}_{n+1}$, the σ-fields $\mathcal{F}_n$ and $\mathcal{T}$ are independent. Being $\mathcal{T}$-measurable, 1_A is therefore independent of $\mathcal{F}_n$ for any n. This means that

$$\xi_n = E\left(1_A | \mathcal{F}_n\right) = E\left(1_A\right) = P(A) \quad \text{a.s.}$$

Therefore the limit $\lim_{n\to\infty} \xi_n = \xi$ is also constant and equal to $P(A)$ a.s. This means that $P(A) = 1_A$ a.s., so $P(A) = 0$ or 1. □

Exercise 4.8

Show that if $A_n \in \sigma\left(\xi_n\right)$ for each n, then the events

$$\limsup_n A_n = \bigcap_{j\geq 1} \bigcup_{i\geq j} A_i$$

and

$$\liminf_n A_n = \bigcup_{j\geq 1} \bigcap_{i\geq j} A_i$$

belong to the tail σ-field $\mathcal{T}$.

Hint You need to write $\limsup_n A_n$ and $\liminf_n A_n$ in terms of $A_k, A_{k+1}, \ldots$ for any k, that is, to show that $\limsup_n A_n$ and $\liminf_n A_n$ will not be affected if any finite number of sets are removed from the sequence $A_1, A_2, \ldots$.

Exercise 4.9

Use Kolmogorov's 0–1 law to show that in a sequence of coin tosses there are a.s. infinitely many heads.

Hint Show that the event

$$\{\eta_1, \eta_2, \ldots \text{ contains infinitely many heads}\}$$

belongs to the σ-field $\mathcal{T}$. Can the probability of this event be 0? The probability of the event

$$\{\eta_1, \eta_2, \ldots \text{ contains infinitely many tails}\}$$

should be the same. Can both probabilities be equal to 0? Can you simultaneously have finitely many heads and finitely many tails in the sequence $\eta_1, \eta_2, \ldots$?

4.4 Solutions

Solution 4.1

Because $\alpha_1 = 0$ is constant, it is $\mathcal{F}_0 = \{\emptyset, \Omega\}$-measurable. Suppose that α_n is $\mathcal{F}_{n-1}$-measurable for some $n = 1, 2, \ldots$. Then

$$\{\alpha_n = 0, \xi_n < a\} \cup \{\alpha_n = 1, \xi_n \leq b\} \in \mathcal{F}_n$$

because $\mathcal{F}_{n-1} \subset \mathcal{F}_n$ and ξ_n is $\mathcal{F}_n$-measurable. This means that

$$\alpha_{n+1} = 1_{\{\alpha_n=0,\xi_n<a\}\cup\{\alpha_n=1,\xi_n\leq b\}}$$

is $\mathcal{F}_n$-measurable. By induction it follows that α_n is $\mathcal{F}_{n-1}$-measurable for each $n = 1, 2, \ldots$, so $\alpha_1, \alpha_2, \ldots$ is a gambling strategy.

Solution 4.2

For a non-negative supermartingale

$$\sup_n E(|\xi_n|) = \sup_n E(\xi_n) \leq E(\xi_1) = E(|\xi_1|) < \infty,$$

since

$$E(\xi_n) \leq E(\xi_1)$$

for each $n = 1, 2, \ldots$. Thus Doob's Martingale Convergence Theorem implies that ξ_n converges a.s. to an integrable limit.

Solution 4.3

Necessity. Suppose that ξ is integrable. It follows that

$$P\{|\xi| < \infty\} = 1.$$

The sequence of random variables $|\xi| 1_{\{|\xi|>M\}}$ indexed by $M = 1, 2, \ldots$ is monotone and

$$|\xi| 1_{\{|\xi|>M\}} \searrow 0 \quad \text{as } M \to \infty$$

on the set $\{|\xi| < \infty\}$, i.e. a.s. By the monotone convergence theorem for integrals

$$\int_{\{|\xi|>M\}} |\xi| \, dP \searrow 0 \quad \text{as } M \to \infty.$$

It follows that for every $\varepsilon > 0$ there exists an $M > 0$ such that

$$\int_{\{|\xi|>M\}} |\xi| \, dP < \varepsilon.$$

Sufficiency. Take $\varepsilon = 1$. There exists an $M > 0$ such that

$$\int_{\{|\xi|>M\}} |\xi| \, dP < 1.$$

Then

$$\begin{aligned} E(|\xi|) &= \int_{\Omega} |\xi| \, dP \\ &= \int_{\{|\xi|>M\}} |\xi| \, dP + \int_{\{|\xi|\leq M\}} |\xi| \, dP \\ &< 1 + MP\{|\xi| \leq M\} \\ &\leq 1 + M < \infty. \end{aligned}$$

Solution 4.4

For any $M > 0$ and any $n > M$ we have

$$(0, \tfrac{1}{n}) = \{\xi_n > M\},$$

so

$$\int_{\{\xi_n>M\}} \xi_n \, dP = \int_{(0,\frac{1}{n})} n \, dP = 1.$$

This means that there is no $M > 0$ such that for all n

$$\int_{\{\xi_n>M\}} \xi_n \, dP < \frac{1}{2}.$$

The sequence ξ_n is not uniformly integrable.

Solution 4.5

In Example 3.4 it was verified that $\xi_n = E(\xi|\mathcal{F}_n)$ is a martingale. Let $\varepsilon > 0$. By Lemma 4.2 there is a $\delta > 0$ such that

$$P(A) < \delta \Longrightarrow \int_A |\xi| \, dP < \varepsilon.$$

Since

$$E(\xi) = E(\xi_n) \geq \int_{\{|\xi_n|>M\}} |\xi_n| \, dP \geq MP\{|\xi_n| > M\},$$

if we take $M > E(\xi)/\delta$, then

$$P\{|\xi_n| > M\} < \delta.$$

Since $\{|\xi_n| > M\} \in \mathcal{F}_n$,

$$\int_{\{|\xi_n|>M\}} |\xi_n| \, dP = \int_{\{|\xi_n|>M\}} |\xi| \, dP < \varepsilon,$$

proving that $\xi_n = E(\xi|\mathcal{F}_n)$ is a uniformly integrable sequence.

Solution 4.6

Because ξ_n is a uniformly integrable sequence, there is an $M > 0$ such that for all n

$$\int_{\{|\xi_n|>M\}} |\xi_n| \, dP < 1.$$

It follows that

$$\begin{aligned} E(|\xi_n|) &= \int_{\{|\xi_n|>M\}} |\xi_n| \, dP + \int_{\{|\xi_n|\leq M\}} |\xi_n| \, dP \\ &< 1 + MP\{|\xi_n| \leq M\} \\ &< 1 + M < \infty \end{aligned}$$

for all n, proving that ξ_n is a bounded sequence in L^1.

Solution 4.7

By Theorem 4.4, $\xi_n = E(a|\mathcal{F}_n)$ a.s. But $E(a|\mathcal{F}_n) = a$ a.s., which proves that $\xi_n = a$ a.s.

Solution 4.8

Observe that

$$\limsup_n A_n = \bigcap_{j\geq k} \bigcup_{i\geq j} A_i$$

for any k. Since

$$A_i \in \sigma(\xi_i) \subset \mathcal{T}_k$$

for every $i \geq k$, it follows that

$$\limsup_n A_n = \bigcap_{j\geq k} \bigcup_{i\geq j} A_i \in \mathcal{T}_k$$

for every k. Therefore

$$\sup_n A_n \in \mathcal{T}.$$

The argument for $\liminf_n A_n$ is similar.

Solution 4.9

Let $\eta_1, \eta_2, \ldots$ be a sequence of coin tosses, i.e. independent random variables with values $+1, -1$ (heads or tails) taken with probability $\frac{1}{2}$ each. Consider the following event:

$$A = \{\eta_1, \eta_2, \ldots \text{ contains infinitely many heads}\}.$$

This event belongs to the tail σ-field $\mathcal{T}$ because

$$A = \limsup_n A_n,$$

where

$$A_n = \{\eta_n = \text{heads}\} \in \sigma(\eta_n)$$

(see Exercise 4.8). Thus, by Kolmogorov's 0–1 law $P(A) = 0$ or 1. However, it cannot be 0 because the event

$$B = \{\eta_1, \eta_2, \ldots \text{ contains infinitely many tails}\}$$

has the same probability by symmetry and $\Omega = A \cup B$ (there must be infinitely many heads or infinitely many tails).

5
Markov Chains

This chapter is concerned with an interesting class of sequences of random variables taking values in a finite or countable set, called the *state space*, and satisfying the so-called *Markov property*. One of the simplest examples is provided by a symmetric random walk ξ_n with values in the set of integers $\mathbb{Z}$. If ξ_n is equal to some $i \in \mathbb{Z}$ at time n, then in the next time instance $n+1$ it will jump either to $i+1$, with probability $\frac{1}{2}$, or to $i-1$, also with probability $\frac{1}{2}$. What makes this model interesting is that the value of ξ_{n+1} at time $n+1$ depends on the past only through the value at time n. This is the Markov property characterizing Markov chains. There are numerous examples of Markov chains, with a multitude of applications.

From the mathematical point of view, Markov chains are both simple and difficult. Their definition and basic properties do not involve any complicated notions or sophisticated mathematics. Yet, any deeper understanding of Markov chains requires quite advanced tools. For example, this is so for problems related to the long-time behaviour of Markov processes. In this chapter we shall try to maintain a balance between the accessibility of exposition and the depth of mathematical results. Various concepts will be introduced. In particular, we shall discuss the classification of states and its relevance to the asymptotic behaviour of transition probabilities. This will turn out to be closely linked to ergodicity and the existence and uniqueness of invariant measures.

5.1 First Examples and Definitions

Example 5.1

In some homes the use of the telephone can become quite a sensitive issue. Suppose that if the phone is free during some period of time, say the nth minute, then with probability p, where $0 < p < 1$, it will be busy during the next minute. If the phone has been busy during the nth minute, it will become free during the next minute with probability q, where $0 < q < 1$. Assume that the phone is free in the 0th minute. We would like to answer the following two questions.

1) What is the probability x_n that the telephone will be free in the nth minute?

2) What is $\lim_{n\to\infty} x_n$, if it exists?

Denote by A_n the event that the phone is free during the nth minute and let $B_n = \Omega \setminus A_n$ be its complement, i.e. the event that the phone is busy during the nth minute. The conditions of the example give us

$$P(B_{n+1}|A_n) = p, \tag{5.1}$$

$$P(A_{n+1}|B_n) = q, \tag{5.2}$$

We also assume that $P(A_0) = 1$, i.e. $x_0 = 1$. Using this notation, we have $x_n = P(A_n)$. Then the total probability formula, see Exercise 1.10, together with (5.1)–(5.2) imply that

$$\begin{aligned} x_{n+1} &= P(A_{n+1}) \\ &= P(A_{n+1}|A_n)P(A_n) + P(A_{n+1}|B_n)P(B_n) \\ &= (1-p)x_n + q(1-x_n) = q + (1-p-q)x_n. \end{aligned} \tag{5.3}$$

It's a bit tricky to find an explicit formula for x_n. To do so we suppose first that the sequence $\{x_n\}$ is convergent, i.e.

$$\lim_{n\to\infty} x_n = x. \tag{5.4}$$

The elementary properties of limits and equation (5.3), i.e. $x_{n+1} = q + (1-p-q)x_n$, yield

$$x = q + (1-p-q)x. \tag{5.5}$$

The unique solution to the last equation is

$$x = \frac{q}{q+p}. \tag{5.6}$$

In particular,

$$\frac{q}{q+p} = q + (1-p-q)\frac{q}{q+p}. \tag{5.7}$$

Subtracting (5.7) from (5.3), we infer that

$$x_{n+1} - \frac{q}{q+p} = (1-p-q)\left(x_n - \frac{q}{q+p}\right). \tag{5.8}$$

Thus, $\{x_n - \frac{q}{q+p}\}$ is a geometric sequence and therefore, for all $n \in \mathbb{N}$,

$$x_n - \frac{q}{q+p} = (1-p-q)^n\left(x_0 - \frac{q}{q+p}\right).$$

Hence, by taking into account the initial condition $x_0 = 1$, we have

$$\begin{aligned} x_n &= \frac{q}{q+p} + \left(x_0 - \frac{q}{q+p}\right)(1-p-q)^n \\ &= \frac{q}{q+p} + \frac{p}{q+p}(1-p-q)^n. \end{aligned} \tag{5.9}$$

Let us point out that although we have used the assumption (5.4) to derive (5.8), the proof of the latter is now complete. Indeed, having proven (5.9), we can show that the assumption (5.4) is indeed satisfied. This is because the conditions $0 < p, q < 1$ imply that $|1-p-q| < 1$, and so $(1-p-q)^n \to 0$ as $n \to \infty$. Thus, (5.4) holds. This provides an answer to the second part of the example, i.e. $\lim_{n\to\infty} x_n = \frac{q}{p+q}$.

The following exercise is a modification of the last example.

Exercise 5.1

In the framework of Example 5.1, let y_n denote the probability that the telephone is busy in the nth minute. Supposing that $y_0 = 1$, find an explicit formula for y_n and, if it exists, $\lim_{n\to\infty} y_n$.

Hint This exercise can be solved directly by repeating the above argument, or indirectly by using some of the results in Example 5.1.

Remark 5.1

The formulae (5.3) and (5.64) can be written collectively in a compact form by using vector and matrix notation. First of all, since $x_n + y_n = 1$, we get

$$\begin{aligned} x_{n+1} &= (1-p)x_n + qy_n, \\ y_{n+1} &= px_n + (1-q)y_n. \end{aligned}$$

Hence, the matrix version takes the form

$$\begin{bmatrix} x_{n+1} \\ y_{n+1} \end{bmatrix} = \begin{bmatrix} 1-p & q \\ p & 1-q \end{bmatrix} \begin{bmatrix} x_n \\ y_n \end{bmatrix}.$$

The situation described in Example 5.1 is quite typical. Often the probability of a certain event at time $n+1$ depends only on what happens at time n, but not further into the past. Example 5.1 provides us with a simple case of a Markov chain. See also the following definition and exercises.

Definition 5.1

Suppose that S is a finite or a countable set. Suppose also that a probability space $(\Omega, \mathcal{F}, P)$ is given. An S-valued sequence of random variables ξ_n, $n \in \mathbb{N}$, is called an S-valued *Markov chain* or a Markov chain on S if for all $n \in \mathbb{N}$ and all $s \in S$

$$P(\xi_{n+1} = s | \xi_0, \ldots, \xi_n) = P(\xi_{n+1} = s | \xi_n). \tag{5.10}$$

Here $P(\xi_{n+1} = s|\xi_n)$ is the conditional probability of the event $\{\xi_{n+1} = s\}$ with respect to random variable ξ_n, or equivalently, with respect to the σ-field $\sigma(\xi_n)$ generated by ξ_n. Similarly, $P(\xi_{n+1} = s|\xi_0, \ldots, \xi_n)$ is the conditional probability of $\{\xi_{n+1} = s\}$ with respect to the σ-field $\sigma(\xi_0, \cdots, \xi_n)$ generated by the random variables $\xi_0, \cdots, \xi_n$.

Property (5.10) will usually be referred to as the *Markov property* of the Markov chain ξ_n, $n \in \mathbb{N}$. The set S is called the *state space* and the elements of S are called *states*.

Proposition 5.1

The model in Example 5.1 and Exercise 5.1 is a Markov chain.

Proof

Let $S = \{0, 1\}$, where 0 and 1 represent the states of the phone being free or busy. First we need to construct an appropriate probability space. Let Ω be the set of all sequences $\omega_0, \omega_1, \ldots$ with values in S. Let μ_0 be any probability measure on S. For example, $\mu = \delta_0$ corresponds to the case when the phone is free at time 0. We shall define P by induction. For any S-valued sequence $s_0, s_1, \ldots$ we put

$$P(\{\omega \in \Omega : \ \omega_0 = s_0\}) = \mu_0(\{s_0\}) \tag{5.11}$$

and

$$P(\{\omega \in \Omega : \omega_i = s_i, \ i = 0, \cdots, n+1\}) \tag{5.12}$$

$$= p(s_{n+1}|s_n)P(\{\omega \in \Omega : \omega_i = s_i,\ i = 0, \cdots, n\}),$$

where $p(s|r)$ are the entries of the 2×2 matrix

$$\begin{bmatrix} p(0|0) & p(0|1) \\ p(1|0) & p(1|1) \end{bmatrix} = \begin{bmatrix} 1-p & q \\ p & 1-q \end{bmatrix}.$$

It seems reasonable to expect P to be a probability measure (with respect to the trivial σ-field of all subsets of Ω). Take this for granted, check only that $P(\Omega) = 1$.

How would you define the process ξ_n, $n \in \mathbb{N}$? We shall do it in the standard way, i.e.

$$\xi_n(\omega) = \omega_n, \quad \omega \in \Omega. \tag{5.13}$$

First we shall show that the transition probabilities of ξ_n are what they should be, i.e.

$$P(\xi_{n+1} = 1|\xi_n = 0) = p, \tag{5.14}$$

$$P(\xi_{n+1} = 0|\xi_n = 1) = q. \tag{5.15}$$

The definition of conditional probability yields

$$P(\xi_{n+1} = 0|\xi_n = 1) = \frac{P(\xi_{n+1} = 0,\ \xi_n = 1)}{P(\xi_n = 1)}.$$

Next, the definition of P gives

$$\begin{aligned}
&P(\xi_{n+1} = 1,\ \xi_n = 0) \\
&= P(\{\omega \in \Omega : \omega_n = 0, \omega_{n+1} = 1\}) \\
&= \sum_{s_0, \cdots, s_{n-1} \in S} P(\{\omega \in \Omega : \omega_i = s_i, i = 0, \cdots, n-1,\ \omega_n = 0, \omega_{n+1} = 1\}) \\
&= \sum_{s_0, \cdots, s_{n-1} \in S} pP(\{\omega \in \Omega : \omega_i = s_i, i = 0, \cdots, n-1,\ \omega_n = 0\}) \\
&= pP(\xi_n = 0),
\end{aligned}$$

by (5.12). We have proven (5.14). Moreover, (5.15) follows by the same argument. A similar line of reasoning shows that ξ_n is indeed a Markov chain. For this we need to verify that for any $n \in \mathbb{N}$ and any $s_0, s_1, \cdots, s_{n+1} \in S$

$$P(\xi_{n+1} = s_{n+1}|\xi_0 = s_0, \cdots, \xi_n = s_n) = P(\xi_{n+1} = s_{n+1}|\xi_n = s_n).$$

We have

$$\begin{aligned}
&P(\xi_0 = s_0, \cdots, \xi_n = s_n, \xi_{n+1} = s_{n+1}) \\
&= P(\{\omega \in \Omega : \omega_i = s_i, i = 0, \cdots, n+1\}) \\
&= p(s_{n+1}|s_n)P(\{\omega \in \Omega : \omega_i = s_i, i = 0, \cdots, n\}) \\
&= p(s_{n+1}|s_n)P(\xi_0 = s_0, \cdots, \xi_n = s_n)
\end{aligned}$$

which, in view of the definition of conditional probability, gives

$$P(\xi_{n+1} = s_{n+1}|\xi_0 = s_0, \cdots, \xi_n = s_n) = p(s_{n+1}|s_n).$$

On the other hand, by an easy generalization of (5.14) and (5.15)

$$P(\xi_{n+1} = s_{n+1}|\xi_n = s_n) = p(s_{n+1}|s_n),$$

which proves (5.10). □

In Example 5.1 the transition probabilities from state i to state j do not depend on the time n. This is an important class of Markov chains.

Definition 5.2

An S-valued Markov chain ξ_n, $n \in \mathbb{N}$, is called *time-homogeneous* or *homogeneous* if for all $n \in \mathbb{N}$ and all $i, j \in S$

$$P(\xi_{n+1} = j|\xi_n = i) = P(\xi_1 = j|\xi_0 = i). \tag{5.16}$$

The number $P(\xi_1 = j|\xi_0 = i)$ is denoted by $p(j|i)$ and called the *transition probability* from state i to state j. The matrix $P = [p(j|i)]_{j,i\in S}$ is called the *transition matrix* of the chain ξ_n.

Exercise 5.2

In the discussion so far we have seen an example of a transition matrix, $P = \begin{bmatrix} 1-p & q \\ p & 1-q \end{bmatrix}$. Obviously the sum of the entries in each column of P is equal to 1. Prove that this is true in general.

Hint Remember that $P(\Omega|A) = 1$ for any event A.

Definition 5.3

$A = [a_{ji}]_{i,j\in S}$ is called a *stochastic matrix* if

1) $a_{ji} \geq 0$, for all $i, j \in S$;

2) the sum of the entries in each column is 1, i.e. $\sum_{j\in S} a_{ji} = 1$ for any $i \in S$.

A is called a *double stochastic matrix* if both A and its transpose A^t are stochastic matrices.

Proposition 5.2

Show that a stochastic matrix is doubly stochastic if and only if the sum of the entries in each row is 1, i.e. $\sum_{i\in S} a_{ji} = 1$ for any $j \in S$.

Proof

Put $A^t = [b_{ij}]$. Then, by the definition of the transposed matrix, $b_{ij} = a_{ji}$. Therefore, A^t is a stochastic matrix if and only if

$$\sum_i a_{ji} = \sum_i b_{ij} = 1,$$

completing the proof. □

Exercise 5.3

Show that if $P = [p_{ji}]_{j,i\in S}$ is a stochastic matrix, then any natural power P^n of P is a stochastic matrix. Is the corresponding result true for a double stochastic matrix?

Hint Show that if A and B are two stochastic matrices, then so is BA. For the second problem, recall that $(BA)^t = A^t B^t$.

Exercise 5.4

Let $P = \begin{bmatrix} 1-p & q \\ p & 1-q \end{bmatrix}$. Show that

$$P^2 = \begin{bmatrix} 1+p^2-2p+pq & 2q-pq-q^2 \\ 2p-pq-p^2 & 1+q^2-2q+pq \end{bmatrix}.$$

Hint This is just simple matrix multiplication.

We see that there is a problem with finding higher powers of the matrix P. When multiplying P^2 by P, P^2, and so on, we obtain more and more complicated expressions.

Definition 5.4

The *n-step transition matrix* of a Markov chain ξ_n with transition probabilities $p(j|i)$, $j, i \in S$ is the matrix P_n with entries

$$p_n(j|i) = P(\xi_n = j|\xi_0 = i). \tag{5.17}$$

Exercise 5.5

Find an exact formula for P_n for the matrix P from Exercise 5.4.

Hint Put $x_n = P(\xi_n = 0|\xi_0 = 0)$ and $y_n = P(\xi_n = 1|\xi_0 = 1)$. Is it correct to suppose that $p_n(0|0) = x_n$ and $p_n(1|1) = y_n$? If yes, you may be able use Example 5.1 and Exercise 5.1.

Exercise 5.6

You may suspect that P_n equals P^n, the nth power of the matrix P. This holds for $n = 1$. Check if it is true for $n = 2$. If this is the case, try to prove that $P_n = P^n$ for all $n \in \mathbb{N}$.

Hint Once again, this is an exercise in matrix multiplication.

The following is a generalization of Exercise 5.6.

Proposition 5.3 (Chapman–Kolmogorov equation)

Suppose that ξ_n, $n \in \mathbb{N}$, is an S-valued Markov chain with n-step transition probabilities $p_n(j|i)$. Then for all $k, n \in \mathbb{N}$

$$p_{n+k}(j|i) = \sum_{s\in S} p_n(j|s)p_k(s|i), \quad i, j \in S. \tag{5.18}$$

Exercise 5.7

Prove Proposition 5.3.

Hint $p_{n+k}(j|i)$ are the entries of the matrix $P_{n+k} = P^{n+k}$.

Proof (of Proposition 5.3)

Let P and P_n be, respectively, the transition probability matrix and the n-step transition probability matrix. Since $p_n(j|i)$ are the entries of P_n, we only need to show that $P_n = P^n$ for all $n \in \mathbb{N}$. This can be done by induction. The assertion is clearly true for $n = 1$. Suppose that $P_n = P^n$. Then, for $i, j \in S$, by the total probability formula and the Markov property (5.10)

$$\begin{aligned} p_{n+1}(j|i) &= P(\xi_{n+1} = j|\xi_0 = i) \\ &= \sum_{s\in S} P(\xi_{n+1} = j|\xi_0 = i, \xi_n = s)P(\xi_n = s|\xi_0 = i) \\ &= \sum_{s\in S} P(\xi_{n+1} = j|\xi_n = s)P(\xi_n = s|\xi_0 = i) \end{aligned}$$

$$= \sum_{s \in S} p(j|s) p_n(s|i),$$

which proves that $P_{n+1} = PP_n$. □

Exercise 5.8 (random walk)

Suppose that $S = \mathbb{Z}$. Let η_n, $n \geq 1$ be a sequence of independent identically distributed random variables with $P(\eta_1 = 1) = p$ and $P(\eta_1 = 0) = q = 1 - p$. Define $\xi_n = \sum_{i=1}^n \eta_i$ for $n \geq 1$ and $\xi_0 = 0$. Show that ξ_n is a Markov chain with transition probabilities

$$p(j|i) = \begin{cases} p, & \text{if } \ j = i + 1, \\ q, & \text{if } \ j = i - 1, \\ 0, & \text{otherwise.} \end{cases}$$

ξ_n, $n \geq 0$, is called a *random walk* starting at 0. Replacing $\eta_0 = 0$ with $\eta_0 = i$, we get a random walk starting at i.

Hint $\xi_{n+1} = \xi_n + \eta_{n+1}$. Are ξ_n and η_{n+1} independent?

Exercise 5.9

For the random walk ξ_n defined in Exercise 5.8 prove that

$$P(\xi_n = j|\xi_0 = i) = \binom{n}{\frac{n+j-i}{2}} p^{\frac{n+j-i}{2}} q^{\frac{n-j+i}{2}} \tag{5.19}$$

if $n + j - i$ is an even non-negative integer, and $P(\xi_n = j|\xi_0 = i) = 0$ otherwise.

Hint Use induction. Note that $\binom{n}{\frac{n+j-i}{2}} p^{\frac{n+j-i}{2}}$ equals 0 if $|j - i| \geq n + 1$.

Proposition 5.4

For all $p \in (0, 1)$

$$P(\xi_n = i|\xi_0 = i) \to 0, \quad \text{as } n \to \infty. \tag{5.20}$$

Proof

To begin with, we shall consider the case $p \neq \frac{1}{2}$. When $j = i$, formula (5.19) becomes

$$P(\xi_n = i|\xi_0 = i) = \begin{cases} \frac{(2k)!}{(k!)^2} (pq)^k, & \text{if } \ n = 2k, \\ 0, & \text{if } \ n \ \text{ is odd.} \end{cases} \tag{5.21}$$

Then, denoting $a_k = \frac{(2k)!}{(k!)^2}(pq)^k$, we have

$$\frac{a_{k+1}}{a_k} = pq\frac{(2k+1)(2k+2)}{(k+1)^2} \to 4pq < 1.$$

Hence, $a_k \to 0$. Thus, $P(\xi_{2k} = i|\xi_0 = i) \to 0$. The result follows, since $P(\xi_{2k+1} = i|\xi_0 = i) = 0 \to 0$.

This argument does not work for $p = \frac{1}{2}$ because $4pq = 1$. In this case we shall need the Stirling formula[1]

$$k! \sim \sqrt{2\pi k}\left(\frac{k}{e}\right)^k, \quad \text{as } k \to \infty. \tag{5.22}$$

Here we use the standard convention: $a_n \sim b_n$ whenever $\frac{a_n}{b_n} \to 1$ as $n \to \infty$. By (5.22)

$$\begin{aligned} a_k &\sim \frac{\sqrt{4\pi k}}{2\pi k}\left(\frac{2k}{e}\right)^{2k}\left(\frac{e}{k}\right)^{2k}(pq)^k \\ &= \frac{1}{\sqrt{\pi k}} \to 0, \quad \text{as} \quad k \to \infty. \end{aligned}$$

Let us note that the second method works in the first case too. However, in the first case there is no need for anything as sophisticated as the Stirling formula. □

Proposition 5.5

The probability that the random walk ξ_n ever returns to the starting point is

$$1 - |p - q|.$$

Proof

Suppose that $\xi_0 = 0$ and denote by $f_0(n)$ the probability that the process returns to 0 at time n for the first time, i.e.

$$f_0(n) = P(\xi_n = 0, \xi_i \neq 0, i = 1, \cdots, n-1).$$

If also $p_n = P(\xi_n = 0)$ for any $n \in \mathbb{N}$, then we can prove that

$$\sum_{n=1}^{\infty} p_0(n) = \sum_{n=0}^{\infty} p_0(n) \sum_{n=1}^{\infty} f_0(n). \tag{5.23}$$

[1] See, for example, E.C. Titchmarsh, *The Theory of Functions*, Oxford University Press, Oxford, 1978.

Since all the numbers involved are non-negative, in order to prove (5.23) we need only to show that

$$p_0(n) = \sum_{k=1}^{n} f_0(k) p_0(n-k) \text{ for } n \geq 1.$$

The total probability formula and the Markov property (5.10) yield

$$\begin{aligned} p_0(n) &= \sum_{k=1}^{n} P(\xi_n = 0, \xi_k = 0, \xi_i \neq 0, i = 1, \cdots, k-1) \\ &= \sum_{k=1}^{n} P(\xi_k = 0, \xi_i \neq 0, i = 1, \cdots, k-1) \\ &\qquad\qquad \times P(\xi_n = 0 | \xi_k = 0, \xi_i \neq 0, i = 1, \cdots, k-1) \\ &= \sum_{k=1}^{n} P(\xi_k = 0, \xi_i \neq 0, i = 1, \cdots, k-1) P(\xi_n = 0 | \xi_k = 0) \\ &= \sum_{k=1}^{n} f_0(k) p_0(n-k). \end{aligned}$$

Having proved (5.23), we are going to make use of it. First we notice that the probability that the process will ever return to 0 equals $\sum_{n=1}^{\infty} f_0(n)$. Next, from (5.23) we infer that

$$\begin{aligned} P(\exists n \geq 1 : \xi_n = 0) &= \sum_{n=1}^{\infty} f_0(n) \\ &= 1 - \left(\sum_{n=0}^{\infty} p_0(n) \right)^{-1} = 1 - \left(\sum_{k=0}^{\infty} p_0(2k) \right)^{-1}. \end{aligned}$$

Since $p_0(2k) = \frac{(2k)!}{(k!)^2} (pq)^k$ and

$$\sum_{k=0}^{\infty} \binom{2k}{k} x^k = (1-4x)^{-1/2}, \ |x| < \frac{1}{4}, \tag{5.24}$$

it follows, that for $p \neq 1/2$

$$P(\exists n \geq 1 : \xi_n = 0) = 1 - (1-4pq)^{1/2} = 1 - |p-q|, \tag{5.25}$$

since, recalling that $q = 1-p$, we have $1 - 4pq = 1 - 4p + 4p^2 = (1-2p)^2 = (q-p)^2$.

The case $p = 1/2$ is more delicate and we shall not pursue this topic here. Let us only remark that the case $p = 1/2$ needs a special treatment as in Proposition 5.4. □

Exercise 5.10

Prove formula (5.24).

Hint Use the Taylor formula to expand the right-hand side of (5.24) into a power series.

Exercise 5.11 (branching process)

On the island Elschenbieden there lives an almost extinct species called Vugiel. Vugiel's males can produce zero, one, two, $\cdots$ male offspring with probability $p_0, p_1, p_2, \cdots$ respectively, where $p_i \geq 0$ and $\sum_{i=0}^{\infty} p_i = 1$. A challenging problem would be to find the Vugiel's chances of survival assuming that each individual lives exactly one year. At this moment, we ask you only to rewrite the problem in the language of Markov chains.

Hint The number of descendants of each male has the same distribution.

Exercise 5.12

Consider the following two cases:

1) In Exercise 5.11 suppose that

$$p_m = \binom{N}{m} p^m (1-p)^{N-m}.$$

for some $p \in (0,1)$ and $N \in \mathbb{N}^*$, where $\mathbb{N}^* = \{1, 2, 3, \cdots\}$. (Note that $p_m = 0$ if $m > N$.) Show that

$$p(j|i) = \binom{Ni}{j} p^j (1-p)^{Ni-j}. \tag{5.26}$$

Deduce that, in particular, $p(j|i) = 0$ if $j > Ni$.

2) Suppose that

$$p_m = \frac{\lambda^m}{m!} e^{-\lambda}, \quad m \in \mathbb{N},$$

for some $\lambda > 0$. In other words, assume that each X_j has the Poisson distribution with mean λ. Show that

$$p(j|i) = \frac{(\lambda i)^j}{j!} e^{-\lambda i}, \quad j \geq i \geq 0. \tag{5.27}$$

Hint If X_1 has the binomial distribution $P(X_1 = m) = \binom{N}{m} p^m (1-p)^{N-m}$, $m \in \mathbb{N}$, then there exists a finite sequence $\eta_1^1, \cdots, \eta_N^1$ of independent identically distributed random variables such that $P(\eta_j^1 = 1) = p$, $P(\eta_j^1 = 0) = 1 - p$ and $X_j = \eta_1^j + \cdots +$

$\eta_N^j = m$. Hence, $X_1 + \cdots + X_i = \sum_{r=1}^{i} \sum_{s=1}^{N} \eta_s^r$, i.e. the sum of Ni independent identically distributed random variables with distribution as above. Hence we infer (5.26).

Proposition 5.6

The probability of survival in Exercise 5.12 equals 0 if $\lambda \le 1$, and $1 - \hat{r}$ when $\lambda > 1$, where $\hat{r} \in (0,1)$ is a solution to

$$r = e^{(r-1)\lambda}. \tag{5.28}$$

Proof

We denote by $\phi(i)$, $i \in \mathbb{N}$ the probability of dying out subject to the condition $\xi_0 = i$. Hence, if $A = \{\xi_n = 0 \text{ for } n \in \mathbb{N}\}$, then

$$\phi(i) = P\left(A | \xi_0 = i\right). \tag{5.29}$$

Obviously, $\phi(0) = 1$ and the total probability formula together with the Markov property (5.10) imply that for each $i \in \mathbb{N}$

$$\begin{aligned} \phi(i) &= \sum_{j=0}^{\infty} P\left(A | \xi_0 = i, \xi_1 = j\right) P\left(\xi_1 = j | \xi_0 = i\right) \\ &= \sum_{j=0}^{\infty} P\left(A | \xi_1 = j\right) P\left(\xi_1 = j | \xi_0 = i\right) \\ &= \sum_{j=0}^{\infty} \phi(j) p(j|i). \end{aligned}$$

Therefore, the sequence $\phi(i)$, $i \in \mathbb{N}$ is bounded (by 1 from above and by 0 from below) and satisfies the following system of equations

$$\begin{aligned} \phi(i) &= \sum_{j=0}^{\infty} \phi(j) p(j|i), \quad i \in \mathbb{N}, \\ \phi(0) &= 1. \end{aligned} \tag{5.30}$$

So far, we have not used any particular distribution of X_j. From now on, we shall assume that the X_j have the Poisson distribution. Hence, by Exercise 5.12, $p(j|i) = \frac{(i\lambda)^j}{j!} e^{-i\lambda}$. It is not an easy problem to find a solution to (5.30), even in this special case. $\phi(i)$ is the probability that the population will die out, subject to the condition that initially there were i individuals. Since we assume that reproduction of different individuals is independent, it is reasonable to make the following *Ansatz*:

$$\phi(i) = A[\phi(1)]^i, \quad i \in \mathbb{N}, \tag{5.31}$$

for some $A > 0$. Although it is possible to prove this Ansatz, we shall not do so here. Note that the boundary condition $\phi(0) = 1$ implies that $A = 1$. Substituting (5.31) (with $A = 1$ and $r := \phi(1)$) into (5.30), we get

$$\begin{aligned} r^i &= \sum_{j=0}^{\infty} r^j \frac{(i\lambda)^j}{j!} e^{-i\lambda} \\ &= e^{-i\lambda} \sum_{j=0}^{\infty} \frac{1}{j!} (ir\lambda)^j = e^{-i\lambda} e^{ir\lambda}. \end{aligned}$$

Hence, r should satisfy

$$r = e^{(r-1)\lambda}. \tag{5.32}$$

Since the function $g(r) = e^{(r-1)\lambda}$, $r \in [0,1]$, is convex, there exist at most two solutions to the equation (5.32). Obviously, one of them is $r = 1$. A bit of analysis, not included here, shows the following:

1) If $\lambda \leq 1$, then the only solution to (5.32) in $[0,1]$ is $r = 1$.

2) If $\lambda > 1$, then there exists a second solution $\hat{r} \in (0,1)$ of the equation (5.32).

In case 1) the situation is simple. We have $\phi(i) = 1$ for all i, and thus the probability of extinction is certain for any initial number of individuals. Case 2) is slightly more involved. The first question we need to address is which of the two solutions of (5.32) gives the correct value of $\phi(1)$? Recall that $p_k = \frac{\lambda^k}{k!} e^{-\lambda}$. Define

$$F(x) = \sum_{k=0}^{\infty} p_k x^k = \sum_{k=0}^{\infty} \frac{\lambda^k}{k!} e^{-\lambda} x^k = e^{\lambda x} e^{-\lambda}, \quad |x| \leq 1. \tag{5.33}$$

Since $P(\xi_1 = 0 | \xi_0 = 1) = p_0$ and

$$\begin{aligned} P(\xi_2 = 0 | \xi_0 = 1) &= \sum_{i=0}^{\infty} P(\xi_2 = 0 | \xi_1 = i) \, P(\xi_1 = i | \xi_0 = 1) \\ &= \sum_{i=0}^{\infty} (p_0)^i p_i = F(p_0) = F(F(0)), \end{aligned}$$

we guess that the following holds:

$$P(\xi_n = 0 | \xi_0 = 1) = F^{(n)}(0), \tag{5.34}$$

where $F^{(n)}$ is the n-fold composition of F. To prove (5.34) it is enough to prove it for n, while assuming it holds for $n - 1$. We have

$$P(\xi_n = 0 | \xi_0 = 1) = \sum_{i=0}^{\infty} P(\xi_n = 0 | \xi_1 = i) \, P(\xi_1 = i | \xi_0 = 1)$$

$$
\begin{aligned}
&= \sum_{i=0}^{\infty} p_i P\left(\xi_{n-1} = 0 | \xi_0 = i\right) \\
&= \sum_{i=0}^{\infty} p_i \left[P(\xi_{n-1} = 0 | \xi_0 = 1)\right]^i \\
&= \sum_{i=0}^{\infty} p_i \left[F^{(n-1)}(0)\right]^i = F(F^{(n-1)}(0)) = F^{(n)}(0).
\end{aligned}
$$

Since the event $\{\xi_n = 0\}$ is contained in events $\{\xi_{n+1} = 0\}$ for all $n \in \mathbb{N}$, we have

$$
\begin{aligned}
\phi(1) &= P\left\{\xi_n = 0, \text{ for some } n \in \mathbb{N} | \xi_0 = 1\right\} \\
&= \lim_{n\to\infty} P\left\{\xi_n = 0 | \xi_0 = 1\right\}
\end{aligned}
$$

by the Lebesgue monotone convergence theorem. Therefore, we infer that

$$
\phi(1) = \lim_{n\to\infty} F^{(n)}(0).
$$

With $F^{(0)}(x) = x$ we only need to show that

$$
F^{(n)}(0) \le \hat{r}, \quad n \in \mathbb{N}. \tag{5.35}
$$

Indeed, once the inequality (5.35) is proven, we infer that $\phi(1) \le \hat{r}$ and thus $\phi(1) = \hat{r}$. We shall prove (5.35) by induction. It is obviously valid for $n = 0$, so we need to study the inductive step. We have

$$
F^{(n)}(0) = F\left(F^{(n-1)}(0)\right) \le F(\hat{r}) = \hat{r},
$$

since F is increasing. We conclude that in the case $\lambda > 1$ the population will become extinct with positive probability.

In the simplest example of the binomial distribution case, i.e. when $N = 1$, equations (5.30) become

$$
\phi(i) = \sum_{j=0}^{i} \phi(j) \binom{i}{j} p^j (1-p)^{i-j}, \quad i \in \mathbb{N}.
$$

Since $\phi(0) = 1$, $\phi(1)$ satisfies

$$
\phi(1) = q + \phi(1)p
$$

with $q = 1 - p$. Hence, trivially, $\phi(1) = 1$. Then, by induction, one proves that $\phi(i) = 1$. Therefore, whatever the initial number of individuals, extinction of the species is certain. □

Remark 5.2

The method presented in the last solution works for any distribution of the variables X_j. It turns out that the mean value λ of X_1 plays the same role as above. One can show that if $\lambda \leq 1$, then the population will become extinct with probability 1, while for $\lambda > 0$ the probability of extinction is larger than 0 and smaller than 1.

Exercise 5.13

On the, now familiar, island of Elschenbieden the question of survival of the Vugiel is a hot political issue. The (human) population of the island is N. Those who believe that action should be taken in order to help the animals preach their conviction quite convincingly. For if a supporter discusses the issue with a non-supporter, the latter will change his mind with probability one. However, they do so only in face-to-face encounters. Suppose that the probability of an encounter of exactly one pair of humans during one day is p and that with probability q this pair is a supporter–non-supporter one. Write down a Markov chain model of this situation. Neglect the probability of two or more encounters during one day.

Hint On each day the number of supporters can either increase by 1 or remain unchanged. What is the probability of the former?

Exercise 5.14 (queuing model)

A car wash machine can serve at most one customer at a time. With probability p, $0 < p < 1$, the machine can finish serving a customer in a unit time. If this happens, the next waiting car (if any) can be served at the beginning of the next unit of time. During the time interval between the nth and $(n+1)$th unit of time the number of cars arriving has the Poisson distribution with parameter $\lambda > 0$. Let ξ_n denote the number of cars being served or waiting to be served at the beginning of unit n. Show that ξ_n, $n \in \mathbb{N}$, is a Markov chain and find its transition probabilities.

Hint Let Z_n, $n = 0, 1, 2, \cdots$ be a sequence of independent identically distributed random variables, each having the Poisson distribution with parameter λ. Then $\xi_{n+1} - \xi_n - Z_n$ equals ± 1 or 0.

Remark 5.3

In the last model we are interested in the behaviour of ξ_n for large values of n. In particular, it is interesting to determine whether the limit of ξ_n or that of $\mathbb{E}\xi_n$ (as $n \to \infty$) exists. In Exercise 5.36 we shall find conditions which guarantee the existence of a unique invariant measure and imply that the Markov chain in question is ergodic.

5.2 Classification of States

In what follows we fix an S-valued Markov chain with transition matrix $P = [p(j|i)]_{j,i\in S}$, where S is a non-empty and at most countable set.

Definition 5.5

A state i is called *recurrent* if the process ξ_n will eventually return to i given that it starts at i, i.e.

$$P(\xi_n = i \text{ for some } n \geq 1|\xi_0 = i) = 1. \tag{5.36}$$

If the condition (5.36) is not satisfied, then the state i is called *transient*.

Theorem 5.1

Show that for a random walk on $\mathbb{Z}$ with parameter $p \in (0,1)$, the state 0 is recurrent if and only if $p = 1/2$. Show that the same holds if 0 is replaced by any other state $i \in \mathbb{Z}$.

Proof

We know from (5.25) that $P(\xi_n = i \text{ for some } n \geq 1|\xi_0 = i) = 1 - |p-q|$ for any $i \in \mathbb{Z}$. □

Definition 5.6

We say that a state i *communicates* with a state j if with positive probability the chain will visit the state j having started at i, i.e.

$$P(\xi_n = j \text{ for some } n \geq 0|\xi_0 = i) > 0. \tag{5.37}$$

If i communicates with j, then we shall write $i \to j$. We say that the state i *intercommunicates* with a state j, and write $i \leftrightarrow j$, if $i \to j$ and $j \to i$.

Exercise 5.15

Show that $i \to j$ if and only if $p_k(j|i) > 0$ for some $k \geq 1$.

Hint Recall that $p_k(j|i) = P(\xi_k = j|\xi_0 = i)$.

Exercise 5.16

Show that

1) $i \leftrightarrow i$,
2) if $i \leftrightarrow j$ then $j \leftrightarrow i$ and
3) if $i \leftrightarrow j$, $j \leftrightarrow k$ then $i \leftrightarrow k$.

In other words, show that $\leftrightarrow$ is an equivalence relation on S.

Hint 1) and 2) are obvious. For 3) use the Chapman–Kolmogorov equations.

Exercise 5.17

For $|x| < 1$ and $j, i \in S$ define

$$P_{ji}(x) = \sum_{n=0}^{\infty} p_n(j|i)x^n, \tag{5.38}$$

$$F_{ji}(x) = \sum_{n=1}^{\infty} f_n(j|i)x^n, \tag{5.39}$$

where $f_n(j|i) = P(\xi_n = j, \xi_k \neq j, k = 1, \cdots, n-1|\xi_0 = i)$. Show that the power series in (5.38)–(5.39) are absolutely convergent for $|x| < 1$ and that

$$P_{ji}(x) = F_{ji}(x)P_{jj}(x), \text{ if } j \neq i, \tag{5.40}$$

$$P_{ii}(x) = 1 + F_{ii}(x)P_{ii}(x). \tag{5.41}$$

Hint Note that $|p_n(j|i)| \leq 1$, so the radius of convergence of the power series (5.38) is ≥ 1.

Exercise 5.18

Show that $\lim_{x \nearrow 1} P_{jj}(x) = \sum_{n=0}^{\infty} p_n(j|j)$ and $\lim_{x \nearrow 1} F_{jj}(x) = \sum_{n=0}^{\infty} f_n(j|j)$.

Hint Apply Abel's lemma[2]: If $a_k \geq 0$ for all $k \geq 0$ and $\limsup_{k\to\infty} \sqrt[k]{|a_k|} \leq 1$, then $\lim_{x \nearrow 1} \sum_{k=0}^{\infty} a_k x^k = \sum_{k=0}^{\infty} a_k$, no matter whether this sum is finite or infinite.

---

[2] For example, see: W. Rudin, *Principles of Mathematical Analysis*, McGraw–Hill Book Company, New York 1976.

Exercise 5.19

Show that a state j is recurrent if and only if $\sum_n p_n(j|j) = \infty$. Deduce that the state j is transient if and only if

$$\sum_n p_n(j|j) < \infty. \tag{5.42}$$

Show that if j is transient, then for each $i \in S$

$$\sum_n p_n(j|i) < \infty. \tag{5.43}$$

Hint If j is recurrent, then $F_{jj}(x) \to \sum_n f_n(j|j) = 1$. Use (5.41) in conjunction with Abel's Lemma.

Exercise 5.20

For a Markov chain ξ_n with transition matrix $P = \begin{bmatrix} 1-p & q \\ p & 1-q \end{bmatrix}$ show that both states are recurrent.

Hint Use Exercise 5.19 and 5.5.

One may suspect that if the state space S is finite, then there must exist at least one recurrent state. For otherwise, if all states were transient and $S = \{1, 2, \cdots, N\}$, then with positive probability a chain starting from 1 would visit 1 only a finite number of times. Thus, after visiting that state for the last time, the chain would move to a different state, say i_2, in which it would stay for a finite time only with positive probability. Thus, in finite time, with positive probability, the chain will never return to states 1 and i_2. By induction, in finite time, with positive probability, the chain will never return to any of the states. This is impossible. The following exercise will give precision to this argument.

Exercise 5.21

Show that if ξ_n is a Markov chain with finite state space S, then there exists at least one recurrent state $i \in S$.

Hint Argue by contradiction and use (5.73).

The following result is quoted here for reference. The proof is surprisingly difficult and falls beyond the scope of this book.

Theorem 5.2

A state $j \in S$ is recurrent if and only if

$$P(\xi_n = j \text{ for infinitely many } n | \xi_0 = j) = 1,$$

and is transient if and only if

$$P(\xi_n = j \text{ for infinitely many } n | \xi_0 = j) = 0.$$

Definition 5.7

For an S-valued Markov chain ξ_n, $n \in \mathbb{N}$, a state $i \in S$ is called *null–recurrent* if it is recurrent and its mean recurrence time m_i defined by

$$m_i := \sum_{n=0}^{\infty} n f_n(i|i) \tag{5.44}$$

equals ∞. A state $i \in S$ is called *positive-recurrent* if it is recurrent and its mean recurrence time m_i is finite.

Remark 5.4

One can show that a recurrent state i is null-recurrent if and only if $p_n(i|i) \to 0$.

We already know that for a random walk on $\mathbb{Z}$ the state 0 is recurrent if and only if $p = 1/2$, i.e. if and only if the random walk is symmetric. In the following problem we shall try to answer if 0 is a null-recurrent or positive-recurrent state (when $p = 1/2$).

Exercise 5.22

Consider a symmetric random walk on $\mathbb{Z}$. Show that 0 is a null-recurrent state. Can you deduce whether other states are positive-recurrent or null-recurrent?

Hint State 0 is null-recurrent if and only if $\sum_n n f_n(0|0) = \infty$. As in Exercise 5.18, $\sum_n n f_n(0|0) = \lim_{x \nearrow 1} F_{00}'(x)$, where F_{00} is defined by (5.39).

Exercise 5.23

For the Markov chain ξ_n from Exercise 5.20 show that not only are all states recurrent, but they are positive-recurrent.

Hint Calculate $f_n(0|0)$ directly.

The last two exercises suggest that the type of a state $i \in S$, i.e. whether it is transient, null-recurrent or positive-recurrent is invariant under the equivalence $\leftrightarrow$. We shall investigate this question in more detail below, but even before doing so we need one more notion: that of a periodic state.

Definition 5.8

Suppose that ξ_n, $n \in \mathbb{N}$, is a Markov chain on a state space S. Let $i \in S$. We say that i is a *periodic state* if and only if the greatest common divisor (gcd) of all $n \in \mathbb{N}^*$, where $\mathbb{N}^* = \{1, 2, 3, \cdots\}$, such that $p_n(i|i) > 0$ is ≥ 2. Otherwise, the state i is called *aperiodic.* In both cases, the gcd is denoted by $d(i)$ and is called the *period* of the state i. Thus, i is periodic if and only if $d(i) \geq 2$. A state i which is positive recurrent and aperiodic is called *ergodic.*

Exercise 5.24

Is this claim that $p_{d(i)}(i|i) > 0$ true or not?

Hint Think of a Markov chain in which it is possible to return to the starting point by two different routes. One route with four steps, the other one with six steps.

One of the by-products of the following exercise is another example of the type asked for in Exercise 5.24.

Exercise 5.25

Consider a Markov chain on $S = \{1, 2\}$ with transition probability matrix $P = \begin{bmatrix} 0 & 1/2 \\ 1 & 1/2 \end{bmatrix}$. This chain can also be described by the graph in Figure 5.1. Find $d(1)$ and $d(2)$.

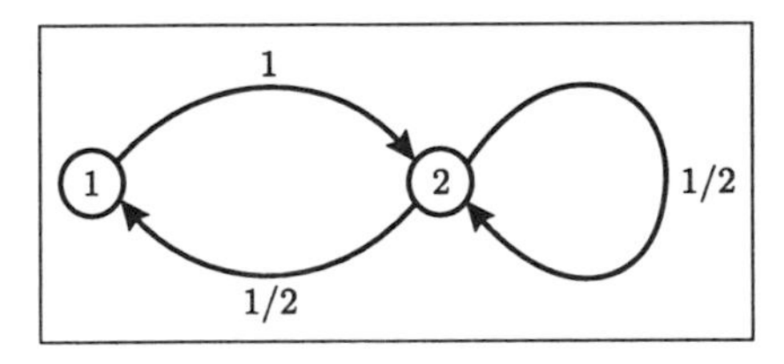

Figure 5.1. Transition probabilities of the Markov chain in Exercise 5.25

Hint Calculate P^2 and P^3. This can be done in two different ways: either algebraically or, probabilistically.

Proposition 5.7

Suppose that $i, j \in S$ and $i \leftrightarrow j$. Show that

1) i is transient if and only if j is;
2) i is recurrent if and only if j is;
3) i is null-recurrent if and only if j is;
4) i is positive-recurrent if and only if j is;
5) i is periodic if and only if j is, in which case $d(i) = d(j)$;
6) i is ergodic if and only if j is.

Proof

It is enough to show properties 1), 4) and 5). Since $i \leftrightarrow j$ one can find $n, m \in \mathbb{N}$ such that $p_m(j|i) > 0$ and $p_n(i|j) > 0$. Hence $\varepsilon := p(j|i)p(i|j)$ is positive. Let us take $k \in \mathbb{N}$. Then by the Chapman–Kolmogorov equations

$$p_{m+k+n}(j|j) = \sum_{r,s \in S} p_m(j|s)p_k(s|r)p_n(r|j) \geq p_m(j|i)p_k(i|i)p_n(i|j) = \varepsilon p_k(i|i).$$

By symmetry

$$p_{n+k+m}(i|i) = \sum_{r,s \in S} p_m(i|s)p_k(s|r)p_n(r|i) \geq p_n(i|j)p_k(j|j)p_m(j|i) = \varepsilon p_k(j|j).$$

Hence, the series $\sum_k p_k(i|i)$ and $\sum_k p_k(j|j)$ are simultaneously convergent or divergent. Hence 1) follows in view of Exercise 5.19.

To prove 5) we use the inequalities derived above. Thus,

$$\begin{aligned} p_{n+k+m}(i|i) &\geq \varepsilon p_k(j|j), \\ p_{n+k+m}(j|j) &\geq \varepsilon p_k(i|i), \end{aligned}$$

for all $k \in \mathbb{N}$.

Put $R(i) = \{k \in \mathbb{N}^* : p_k(i|i) > 0\}$. Then $d(i) = \gcd R(i)$ by definition. Denote by $r(i)$ the minimum of $R(i)$. Then, since the algebraic sum $R(i) + R(i)$ is contained in $R(i)$, we infer that

$$R(i) = r(i) + \mathbb{N}d(i).$$

Put $\lambda = m + n$. The earlier calculations show that $p_{k+\lambda}(j|j) \leq p_k(i|i)$. Hence $R(i) + \lambda \subset R(j)$. Therefore $d(i)|d(j)$, since $R(j) = r(j) + \mathbb{N}d(j)$. By symmetry, $d(j)|d(i)$, which proves that $d(i) = d(j)$. □

Exercise 5.26

Show that the following modification of 2) above is true: If i is recurrent and $i \to j$, then $j \to i$. Deduce that if i is recurrent and $i \to j$, then j is recurrent and $j \leftrightarrow i$.

Hint Is it possible for a chain starting from i to visit j and then never return to i? Is such a situation possible when i is a recurrent state?

The following result describes how the state space S can be partitioned into a countable sum of classes. One of these classes consists of all transient states. Each of the other class consists of interconnecting recurrent states. If the chain enters one of the classes of second type, it will never leave it. However, if the chain enters the class of transient states, it will eventually leave it (and so never return to it). We begin with a definition.

Definition 5.9

Suppose that ξ_n, $n \in \mathbb{N}$, is a Markov chain on a countable state space S.

1) A set $C \subset S$ is called *closed* if once the chain enters C it will never leave it, i.e.

$$P(\xi_k \in S \setminus C \text{ for some } k \geq n | \xi_n \in C) = 0. \tag{5.45}$$

2) A set $C \subset S$ is called *irreducible* if any two elements i, j of C intercommunicate, i.e. $p(j|i) > 0$ for all $i, j \in C$.

Theorem 5.3

Suppose that ξ_n, $n \in \mathbb{N}$, is a Markov chain on a countable state space S. Then

$$S = T \cup \bigcup_{j=1}^{N} C_j, \quad \text{(disjoint sum)}, \tag{5.46}$$

where T is the set of all transient states in S and each C_j is a closed irreducible set of recurrent states.

Exercise 5.27

Suppose that ξ_n, $n \in \mathbb{N}$, is a Markov chain on a countable state space S. Show that a set $C \subset S$ is closed if and only if $p(j|i) = 0$ for all $i \in C$ and $j \in S \setminus C$.

Hint One implication is trivial. For the other one use the countable additivity of the measure P.

Proof (of Theorem 5.3)

Let $R = S \setminus C$ denote the set of all recurrent states. If $i \leftrightarrow j$, the both i and j belong either to T or to R. It follows that the interconnection relation $\leftrightarrow$ restricted to R is an equivalence relation as well. Therefore, $R = \bigcup_{j=1}^{N} C_j$, $C_j = [s_j]$, $s_j \in R$. Here N denotes the number of different equivalent classes. Since by definition each C_j is an irreducible set, we only need to show that it is closed. But this follows from Exercise 5.26. Indeed, if $i \in C_k$ and $i \to j$, then $i \leftrightarrow j$, and so $j \in C_k$. □

5.3 Long-Time Behaviour of Markov Chains: General Case

For convenience we shall denote the countable state space S by $\{1, 2, 3, \cdots\}$ when S is an infinite set and by $\{1, 2, \cdots, n\}$ when S is finite.

Proposition 5.8

Let $P = [p(j|i)]$ be the transition matrix of a Markov chain with state space S. Suppose that for all $i, j \in S$

$$\lim_{n\to\infty} p_n(j|i) =: \pi_j. \tag{5.47}$$

(In particular, the limit is independent of i.) Then

1) $\sum_j \pi_j \leq 1$;
2) $\sum_i p(j|i)\pi_i = \pi_j$;
3) either $\sum_j \pi_j = 1$, or $\pi_j = 0$ for all $j \in S$.

Proof

To begin with, let us assume that S is finite with m elements. Using the Chapman–Kolmogorov equations (5.18), we have

$$\begin{aligned}\sum_{j\in S} \pi_j &= \sum_{j=1}^{m} \pi_j = \sum_{j=1}^{m} \lim_{n\to\infty} p_n(j|i) \\ &= \lim_{n\to\infty} \sum_{j=1}^{m} p_n(j|i) = \lim_{n\to\infty} 1 = 1,\end{aligned}$$

since $\sum_{j=1}^{m} p_n(j|i) = 1$ for any $n \in \mathbb{N}$ (see Exercise 5.2). This proves 1) and 3) simultaneously. Moreover, it shows that the second alternative in 3) can never occur. To prove 2) we argue in a similar way. Let us fix $j \in S$ and (an auxiliary) $k \in S$. Then,

$$\begin{aligned}\sum_{i=1}^{m} p(j|i)\pi_i &= \sum_{i=1}^{m} \lim_{n\to\infty} p(j|i)p_n(i|k) \\ &= \lim_{n\to\infty} \sum_{i=1}^{m} p(j|i)p_n(i|k) = \lim_{n\to\infty} p_{n+1}(j|k) = \pi_j,\end{aligned}$$

since $\sum_{i=1}^{m} p(j|i)p_n(i|k) = p_{n+1}(j|k)$ by the Chapman–Kolmogorov equations.

When the set S is infinite, we cannot just repeat the above argument. The reason is quite simple: in general the two operations lim and $\sum$ cannot be interchanged. They can when the sum is finite, and we used this fact above. But if S *is* infinite, then the situation is more subtle. One possible solution of the difficulty is contained in the following version of the Fatou lemma.

Lemma 5.1 (Fatou)

Suppose that $a_j(n) \geq 0$ for $j, n \in \mathbb{N}$. Then

$$\sum_j \liminf_n a_j(n) \leq \liminf_n \sum_j a_j(n). \tag{5.48}$$

Moreover, if $a_j(n) \leq b_j$ for $j, n \in \mathbb{N}$ and $\sum_j b_j < \infty$, then

$$\limsup_n \sum_j a_j(n) \leq \sum_j \limsup_n a_j(n). \tag{5.49}$$

Using the fact that for a convergent sequence lim and lim inf coincide, by the Fatou lemma we have

$$\begin{aligned}\sum_{j\in S} \pi_j &= \sum_{j=1}^{\infty} \pi_j = \sum_{j=1}^{\infty} \lim_{n\to\infty} p_n(j|i) \\ &\leq \liminf_{n\to\infty} \sum_{j=1}^{\infty} p_n(j|i) = \liminf_{n\to\infty} 1 = 1\end{aligned}$$

since, as before, $\sum_{j=1}^{\infty} p_n(j|i) = 1$ for any $n \in \mathbb{N}$. This proves 1). A similar argument shows 2). Indeed, with $j \in S$ and $k \in S$ fixed, by the Chapman–Kolmogorov equations and the Fatou lemma we have

$$\sum_{i=1}^{\infty} p(j|i)\pi_i = \sum_{i=1}^{\infty} \lim_{n\to\infty} p(j|i)p_n(i|k)$$

$$\leq \liminf_{n\to\infty} \sum_{i=1}^{\infty} p(j|i)p_n(i|k) = \liminf_{n\to\infty} p_{n+1}(j|k) = \pi_j.$$

To complete the proof of 2) suppose that for some $k \in S$

$$\sum_{i=1}^{\infty} p(k|i)\pi_i < \pi_k.$$

Then, since $\sum_{j\in S} \pi_j = \sum_{j\neq k} \pi_j + \pi_k$, by the part of 2) already proven we have

$$\begin{aligned}
\sum_{j=1}^{\infty} \pi_j &> \sum_{j\neq k} \left(\sum_i p(j|i)\pi_i \right) + \sum_i p(k|i)\pi_i \\
&= \sum_{j=1}^{\infty} \sum_i^{\infty} p(j|i)\pi_i = \sum_{i=1}^{\infty} \sum_j^{\infty} p(j|i)\pi_i \\
&= \sum_{i=1}^{\infty} \pi_i \left(\sum_j^{\infty} p(j|i) \right) = \sum_{i=1}^{\infty} \pi_i.
\end{aligned}$$

We used the fact that $\sum_j^{\infty} p(j|i) = 1$ together with (5.65). This contradiction proves 2).

In order to verify 3) observe that by iterating 2) we obtain

$$\sum_i p_n(j|i)\pi_i = \pi_j.$$

Hence,

$$\begin{aligned}
\pi_j &= \lim_{n\to\infty} \sum_i p_n(j|i)\pi_i \\
\sum_i \lim_{n\to\infty} p_n(j|i)\pi_i &= \sum_i \pi_j \pi_i = \pi_j \sum_i \pi_i.
\end{aligned}$$

Therefore, the product $\pi_j \left(\sum_i \pi_i - 1\right)$ is equal to 0 for all $j \in S$. As a result, 3) follows. Indeed, if $\sum_i \pi_i \neq 1$, then $\pi_j = 0$ for all $j \in S$. □

Definition 5.10

A probability measure $\mu := \sum_{j\in S} \mu_j \delta_j$ is an *invariant measure* of a Markov chain ξ_n, $n \in \mathbb{N}$, with transition probability matrix $P = [p(j|i)]$ if for all $n \in \mathbb{N}$ and all $j \in S$

$$\sum_{i\in S} p_n(j|i)\mu_i = \mu_j.$$

Exercise 5.28

Under the assumptions of Proposition 5.8 show that if $\sum_{j\in S}\pi_j = 1$, then $\mu := \sum_{j\in S}\pi_j\delta_j$ is the unique invariant measure of our Markov chain. Here δ_j is the Dirac delta measure at j.

Hint μ is an invariant measure of a Markov chain ξ_n, $n \in \mathbb{N}$, if and only if for each $n \in \mathbb{N}$, the distribution of ξ_n equals μ, provided the same holds for ξ_0.

Exercise 5.29

Show that if $\pi_j = 0$ for all $j \in S$, then there is no invariant measure.

Hint Look closely at the uniqueness part of the solution to Exercise 5.28.

The following exercise shows that a unique invariant measure may exist, even though the condition (5.47) is not satisfied.

Exercise 5.30

Find all invariant measures for a Markov chain whose graph is given in Figure 5.2.

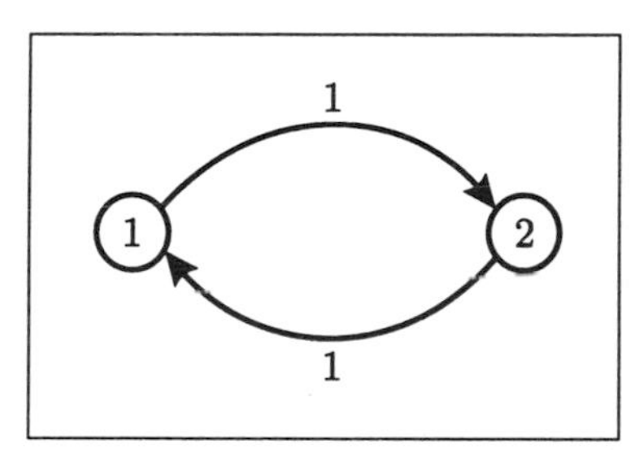

Figure 5.2. Transition probabilities of the Markov chain in Exercise 5.30

Hint Find the transition probability matrix P and solve the vector equation $P\pi = \pi$ for $\pi = (\pi_1, \pi_2)$, subject to the condition $\pi_1 + \pi_2 = 1$.

We shall study some general properties of invariant measures. Above we have seen examples of Markov chains with a unique invariant measure. In what follows we shall investigate the structure of the set of all invariant measures.

Exercise 5.31

Show that if μ and ν are invariant measures and $\theta \in [0,1]$, then $(1-\theta)\mu + \theta\nu$ is also an invariant measure.

Hint Apply the definition of an invariant measure.

Exercise 5.32

Show that if μ is an invariant measure of a Markov chain ξ_n, $n \in \mathbb{N}$ with state space S, then $\operatorname{supp} \mu \subset S \setminus T$, where T denotes (as usual) the set of all transient states.

Hint If j is a transient state, then $p_n(j|i) \to 0$ for all $i \in S$.

The above result shows that there is a close relationship between invariant measures and recurrent states. Below we shall present without proof a couple of results on the existence of such measures and their properties.

Theorem 5.4

Suppose that ξ_n, $n \in \mathbb{N}$, is a Markov chain on a state space $S = T \cup C$, where T is the set of all transient states and C is a closed irreducible set of recurrent states.[3] Then there exists an invariant measure if and only if each element of C is positive-recurrent. Moreover, if this is the case, then the invariant measure is unique and it is given by $\mu = \sum_i \mu_i \delta_i$, where

$$\mu_i = \frac{1}{m_i}$$

with m_i being the mean recurrence time of the state i, see (5.44).

Note, that by Exercise 5.32, the unique invariant measure in Theorem 5.4 is supported by C.

Remark 5.5

If $C = \bigcup_{j=1}^{N} C_j$, where each C_j is a closed irreducible set of recurrent states, then the above result holds, except for the uniqueness part. In fact, if each element of some C_j is positive-recurrent, then there exists a invariant measure μ_j supported by C_j. Moreover, μ_j is the unique invariant measure with support in C_j. In the special case when each element of C is positive-recurrent, every invariant measure μ is a convex combination of the invariant measures μ_j, $j \in \{1, \cdots, N\}$.

Theorem 5.5

Suppose that ξ_n, $n \in \mathbb{N}$, is a Markov chain with state space S. Let $j \in S$ be a recurrent state.

[3] Hence in the decomposition (5.46) the number N of different classes of recurrent states is equal to 1.

1) If j is aperiodic, then

$$p_n(j|j) \to \frac{1}{m_j}. \tag{5.50}$$

Moreover, for any $i \in S$,

$$p_n(j|i) \to \frac{F_{ji}(1)}{m_j}, \tag{5.51}$$

where $F_{ji}(x)$ is defined in (5.39);

2) If j is a periodic state of period $d \geq 2$, then

$$p_{nd}(j|j) \to \frac{d}{m_j}. \tag{5.52}$$

Exercise 5.33

Suppose that ξ_n, $n \in \mathbb{N}$, is a Markov chain with state space S. Let $j \in S$ be a transient state. Show that for any $i \in S$

$$p_n(j|i) \to 0. \tag{5.53}$$

Hint Use Exercise 5.19.

Definition 5.11

A Markov chain ξ_n, $n \in \mathbb{N}$, with state space S is called *ergodic* if each $i \in S$ is ergodic.

Exercise 5.34

Show that if ξ_n, $n \in \mathbb{N}$, is an ergodic Markov chain with state space S, then $p_n(j|i) \to \pi_j$ as $n \to \infty$ for any $j, i \in S$, where $\pi = \sum_i \pi_i \delta_i$ is the unique invariant measure.

Hint Use Theorem 5.5.

Exercise 5.35

Use the last result to investigate whether the random walk on $\mathbb{Z}$ has an invariant measure.

Hint Use Exercise 5.9.

Below we shall see that a converse result to Theorem 5.5 is also true.

Theorem 5.6

Suppose that ξ_n, $n \in \mathbb{N}$, is an irreducible aperiodic Markov chain with state space S. Then ξ_n, $n \in \mathbb{N}$, is ergodic if and only if it has a unique invariant measure.

Proof

The 'if' part is proved in Exercise 5.34. We shall deal with the 'only if' part. Suppose that $\pi = \sum_j \pi_j \delta_j$ is the unique invariant measure of the chain. Then $\pi_j > 0$ for some $j \in S$. Recall that due to Theorem 5.5 and the Exercise 5.33, $\lim_{n\to\infty} p_n(j|i)$ exists for all $i, j \in S$.

Since $\sum_i p_n(j|i)\pi_i = \pi_j$, by the Fatou lemma (inequality (5.49))

$$\sum_i \lim_{n\to\infty} p_n(j|i)\pi_i \geq \limsup_{n\to\infty} \sum \sum_i p_n(j|i)\pi_i = \pi_j.$$

Hence, there exists an $i \in S$ such that $\lim_{n\to\infty} p_n(j|i)\pi_i > 0$. Therefore $\lim_{n\to\infty} p_n(j|i) > 0$, which in view of Theorem 5.5 implies that $m_j < \infty$. Thus, j is an ergodic state and, since the chain is irreducible, all states are ergodic as well. □

Exercise 5.36

Prove that for the Markov chain in Exercise 5.14 there exists an invariant measure if and only if $\lambda' := \sum_{j=0}^{\infty} jq'_j < 1$. Show that if this is the case, then the invariant measure is unique. Conclude that the chain is ergodic if and only if $\lambda' < 1$.

Hint Difficult. If you don't know where to start, read the solution.

5.4 Long-Time Behaviour of Markov Chains with Finite State Space

As we have seen above, the existence of the π_j plays a very important role in the study of invariant measures. In what follows we shall investigate this question in the case when the state space S is finite.

Theorem 5.7

Suppose that S is finite and the transition matrix $P = [p(j|i)]$ of a Markov chain on S satisfies the condition

$$\exists n_0 \in \mathbb{N}\ \exists \varepsilon > 0 : p_{n_0}(j|i) \geq \varepsilon, \quad i, j \in S. \tag{5.54}$$

Then, the following limit exists for all $i, j \in S$ and is independent of i:

$$\lim_{n\to\infty} p_n(j|i) = \pi_j. \tag{5.55}$$

The numbers π_j satisfy

$$\pi_j > 0, j \in S \text{ and } \sum_{j\in S} \pi_j = 1. \tag{5.56}$$

Conversely, if a sequence of numbers π_j, $j \in S$ satisfies conditions (5.55)–(5.56), then assumption (5.54) is also satisfied.

Proof

Denote the matrix $P^{n_0} = [p_{n_0}(j|i)]$ by $Q = [q(j|i)]$. Then the process $\eta_k = \xi_{kn_0}$, $k \in \mathbb{N}$, is a Markov chain on S with transition probability matrix Q satisfying (5.54) with n_0 equal to 1. Note that $p_{kn_0}(j|i) = q_n(j|i)$ due to the Chapman–Kolmogorov equations. Suppose that the properties (5.55)–(5.56) hold true for Q. In particular, $\lim_{k\to\infty} p_{kn_0}(j|i) = \pi_j$ exists and is independent of i. We claim that they are also true for the original matrix P. Obviously, one only needs to check condition (5.55). The Chapman–Kolmogorov equations (and the fact that S is finite) imply that for any $r = 1, \cdots, n_0 - 1$

$$\begin{aligned} p_{kn_0+r}(j|i) &= \sum_{s\in S} p_{kn_0}(j|s) p_r(s|i) \to \sum_{s\in S} \pi_j p_r(s|i) \\ &= \pi_j \sum_{s\in S} p_r(s|i) = \pi_j. \end{aligned}$$

Therefore, by a simple result in calculus, according to which, if for a sequence a_n, $n \in \mathbb{N}$, there exists a natural number n_0 such that for each $r \in \{0, 1, \cdots, n_0 - 1\}$ the limit $\lim_{k\to\infty} a_{kn_0+r}$ exists and is r-independent, then the sequence a_n is convergent to the common limit of those subsequences, we infer that (5.55) is satisfied.

In what follows we shall assume that (5.54) holds with $n_0 = 1$. Let us put $p_0(j|i) = \delta_{ji}$ and for $j \in S$

$$\begin{aligned} m_n(j) &:= \min_{i\in S} p_n(j|i) \\ M_n(j) &:= \max_{i\in S} p_n(j|i) \end{aligned}$$

Observe that $M_0(j) = 1$ and $m_0(j) = 0$ for all $j \in S$. From the Chapman–Kolmogorov equations it follows that the sequence $M_n(j)$, $n \in \mathbb{N}$, is decreasing, while the sequence $m_n(j)$, $n \in \mathbb{N}$, is increasing. Indeed, since $\sum_k p(k|i) = 1$,

$$\begin{aligned} p_{n+1}(j|i) &= \sum_{k\in S} p_n(j|k)p(k|i) \\ &\geq \min_k p_n(j|k) \sum_{k\in S} p(k|i) \\ &= \max_k p_n(j|k) = m_n(j). \end{aligned}$$

Hence, by taking the minimum over all $i \in S$, we arrive at

$$m_{n+1}(j) = \min_i p_n(j|i) \geq m_n(j).$$

Similarly,

$$\begin{aligned} p_{n+1}(j|i) &= \sum_{k\in S} p_n(j|k)p(k|i) \\ &\leq \max_k p_n(j|k) \sum_{k\in S} p(k|i) \\ &= \min_k p_n(j|k) = M_n(j). \end{aligned}$$

Hence, by taking the maximum over all $i \in S$, we obtain

$$M_{n+1}(j) = \max_i p_n(j|i) \leq M_n(j).$$

Since $M_n(j) \geq m_j(n)$, the sequences $M_n(j)$ and $m_n(j)$ are bounded from below and from above, respectively. As a consequence, they both have limits. To show that the limits coincide we shall prove that

$$\lim_{n\to\infty} (M_n(j) - m_n(j)) = 0. \tag{5.57}$$

For $n \geq 0$ we have

$$\begin{aligned} p_{n+1}(j|i) &= \sum_{s\in S} p_n(j|s)p(s|i) \qquad (5.58) \\ &= \sum_{s\in S} p_n(j|s) \left[p(s|i) - \varepsilon p_n(s|j)\right] + \varepsilon \sum_{s\in S} p_n(j|s)\varepsilon p_n(s|j) \\ &= \sum_{s\in S} p_n(j|s) \left[p(s|i) - \varepsilon p_n(s|j)\right] + \varepsilon p_{2n}(j|j) \end{aligned}$$

Chapman–Kolmogorov equations. The expression in square brackets is ≥ 0. Indeed, by assumption (5.54), $p_k(s|i) \geq \varepsilon$ and $p_n(s|j) \leq 1$. Therefore,

$$p_{n+1}(j|i) \geq \min_{s\in S} p_n(j|s) \sum_{s\in S} [p(s|i) - \varepsilon p_n(s|j)] + \varepsilon p_{2n}(j|j) \tag{5.59}$$
$$= m_n(j) \sum_{s\in S} (1-\varepsilon) + \varepsilon p_{2n}(j|j).$$

By taking the minimum over $i \in S$, we arrive at

$$m_{n+1}(j) \geq (1-\varepsilon) m_n(j) + \varepsilon p_{2n}(j|j). \tag{5.60}$$

Recycling the above argument, we obtain a similar inequality for the sequence $M_n(j)$:

$$M_{n+1}(j) \leq (1-\varepsilon) M_n(j) + \varepsilon p_{2n}(j|j). \tag{5.61}$$

Thus, by subtracting (5.60) from (5.61) we get

$$M_{n+1}(j) - m_{n+1} \leq (1-\varepsilon)\left(M_n(j) - m_n(j)\right). \tag{5.62}$$

Hence, by induction

$$M_n(j) - m_n \leq (1-\varepsilon)^n, \quad n \in \mathbb{N}.$$

This proves (5.57). Denote by π_j the common limit of $M_n(j)$ and $m_n(j)$. Then (5.55) follows from (5.57). Indeed, if $i, j \in S$, then

$$m_n(j) \leq p_n(j|i) \leq M_n(j).$$

To prove that $\pi_j > 0$ let us recall that $m_n(j)$ is an increasing sequence and $m_1(j) \geq \varepsilon$ by (5.54). We infer that $\pi_j \geq \varepsilon$. □

Exercise 5.37

Show that $p_n(j|i) \to \pi_j$ at an exponential rate.

Hint Recall that $m_n(j) \leq \pi_j \leq M_n(j)$ and use (5.62).

The above proves the following important result.

Theorem 5.8

Suppose that the transition matrix $P = [p(j|i)]$ of a Markov chain ξ_n, $n \in \mathbb{N}$, satisfies assumption (5.54). Show that there exists a unique invariant measure μ. Moreover, for some $A > 0$, and $\alpha < 1$

$$|p_n(j|i) - \pi_j| \leq A\alpha^n, \quad i, j \in S, n \in \mathbb{N}. \tag{5.63}$$

Proof (of the converse part of Theorem 5.7)

Put

$$\varepsilon = \frac{1}{2} \min_j \pi_j.$$

Since $p_n(j|i) \to \pi_j$ for all $i, j \in S$, there is an $n_0 \in \mathbb{N}$, such that $p_k(j|i) \geq \varepsilon$ for all for $k \geq n_0$ and $(i, j) \in S^2$. Putting $k = n_0$ proves that (5.54) is satisfied.

Let us observe that we have used only two facts: $\pi_j > 0$ for all $j \in S$, and $p_n(j|i) \to \pi_j$ for all $i, j \in S$. □

Exercise 5.38

Investigate the existence and uniqueness of an invariant measure for the Markov chain in Proposition 5.1.

Hint Are the assumptions of Theorem 5.7 satisfied?

Remark 5.6

The solution to Exercise 5.38 allows us to find the unique invariant measure by direct methods, i.e. by solving the linear equations (5.79)–(5.80).

Exercise 5.39

Find the invariant measure from Exercise 5.38 by calculating the limits (5.55).

Hint Refer to Solution 5.5.

Exercise 5.40

Ian plays a fair game of dice. After the nth roll of a die he writes down the maximum outcome ξ_n obtained so far. Show that ξ_n is a Markov chain and find its transition probabilities.

Hint $\xi_{n+1} = \max\{\xi_n, X_{n+1}\}$, where X_k is the outcome of the kth roll.

Exercise 5.41

Analyse the Markov chain described in Exercise 5.40, but with fair die replaced by a fair pyramid.

Hint A pyramid has four faces only.

Exercise 5.42

Suppose that $a \geq 1$ is a natural number. Consider a random walk on $S = \{0, 1, \cdots, a\}$ with absorbing barriers at 0 and a, and with probability p of moving to the right and probability $q = 1 - p$ of moving to the left from any of the states $1, \ldots, a-1$. Hence, our random walk is a Markov chain with transition probabilities

$$p(j|i) = \begin{cases} p & \text{if } 1 \leq i \leq a-1, j = i+1, \\ q & \text{if } 1 \leq i \leq a-1, j = i-1, \\ 1 & \text{if } i = j = 0 \quad \text{or } i = j = a, \\ 0 & \text{otherwise.} \end{cases}$$

Find, (a) all invariant measures (there may be just one), (b) the probability of hitting the right-hand barrier prior to hitting the left-hand one.

Hint For (a) recall Exercise 5.30 and for (b) Exercise 5.12.

5.5 Solutions

Solution 5.1

First we give a direct solution. With A_n and B_n being the events that the phone is free or busy in the nth minute, we have $y_n = P(B_n)$. The total probability formula then yields, for $n \in \mathbb{N}$,

$$P(B_{n+1}) = P(B_{n+1}|A_n)P(A_n) + P(B_{n+1}|B_n)P(B_n)$$

i.e.

$$y_{n+1} = p + (1 - p - q)y_n. \tag{5.64}$$

As before, assuming for the time being that $y = \lim_n y_n$ exists, we find that $y = p + (1-p-q)y$, and so $y = \frac{p}{p+q}$. In particular, $\frac{p}{p+q} = p + (1-p-q)\frac{p}{p+q}$. Subtracting the last equality from (5.64), we see that $\{y_n - \frac{p}{p+q}\}$ is a geometric sequence, so that $y_n - \frac{p}{p+q} = \left(y_0 - \frac{p}{p+q}\right)(1-p-q)^n$. Since $y_0 = 1$, some simple algebra leads to the formula

$$y_n = \frac{p}{p+q} + \frac{q}{p+q}(1-p-q)^n, \quad n \in \mathbb{N}.$$

The last formula can be used to prove that the $\lim_n y_n$ exists and is equal to $\frac{p}{p+q}$.

Another approach is to use the results of Example 5.1. Since $x_0 = 1 - y_0$, by (5.8) we have

$$\begin{aligned} y_n &= 1 - x_n = 1 - \frac{q}{p+q} - \left(1 - y_0 - \frac{q}{p+q}\right)(1-p-q)^n \\ &= \frac{p}{p+q} + \frac{q}{p+q}(1-p-q)^n, \end{aligned}$$

which agrees with the first method of solution.

Solution 5.2

We have to show that $\sum_{j\in S} p(j|i) = 1$ for every $i \in S$. We have

$$\begin{aligned} \sum_{j\in S} p(j|i) &= \sum_{j\in S} P(\xi_1 = j|\xi_0 = i) \\ &= P\left(\cup_{j\in S}\{\xi_1 = j\}|\xi_0 = i\right) = P(\xi_1 \in S|\xi_0 = i) \\ &= P(\Omega|\xi_0 = i) = 1. \end{aligned}$$

Solution 5.3

Suppose that $A = [a_{ji}]_{j,i\in S}$ and $B = [b_{ji}]_{j,i\in S}$ are two stochastic matrices. If $C = BA$, then $c_{ji} = \sum_k b_{jk}a_{ki}$. Hence, for any $i \in S$

$$\begin{aligned} \sum_j c_{ji} &= \sum_j \left(\sum_k b_{jk}a_{ki}\right) = \sum_k \left(\sum_j b_{jk}a_{ki}\right) \\ &= \sum_k \left(\sum_j b_{jk}\right) a_{ki} = \sum_k a_{ki} = 1, \end{aligned}$$

where the last two equalities hold because B and A are stochastic matrices. We have used the well-known fact that

$$\sum_i \sum_j a_{ij} = \sum_j \sum_i a_{ij}, \tag{5.65}$$

for any non-negative double sequence $(a_{ij})_{i,j=1}^{\infty}$ (see, for example, Rudin's book cited in the hint to Exercise 5.18). The above argument implies that P^2 is a stochastic matrix. The desired result follows by induction.

To prove that P^n is a double stochastic matrix whenever P is, it is enough to observe that AB is a double stochastic matrix if A and B are. The latter follows because $(AB)^t = B^t A^t$.

Solution 5.4

Some simple algebra gives

$$\begin{aligned} P^2 &= \begin{bmatrix} (1-p)(1-p)+qp & (1-p)q+q(1-q) \\ p(1-p)+(1-q)p & pq+(1-q)(1-q) \end{bmatrix} \\ &= \begin{bmatrix} 1+p^2-2p+pq & 2q-pq-q^2 \\ 2p-pq-p^2 & 1+q^2-2q+pq \end{bmatrix}. \end{aligned}$$

Solution 5.5

Put $x_n = P(\xi_n = 0|\xi_0 = 0)$ and $y_n = P(\xi_n = 1|\xi_0 = 1)$. We have calculated the formulae for x_n and y_n in Example 5.1 and Exercise 5.1. Since also

$$\begin{aligned} 1-x_n &= \frac{p}{q+p} - \frac{p}{q+p}(1-p-q)^n, \\ 1-y_n &= \frac{q}{q+p} - \frac{q}{q+p}(1-p-q)^n, \end{aligned}$$

we arrive at the following formula for the n-step transition matrix:

$$P_n = \begin{bmatrix} \frac{q}{q+p} + \frac{p}{q+p}(1-p-q)^n & \frac{q}{q+p} - \frac{q}{q+p}(1-p-q)^n \\ \frac{p}{q+p} - \frac{p}{q+p}(1-p-q)^n & \frac{p}{q+p} + \frac{q}{q+p}(1-p-q)^n \end{bmatrix}. \tag{5.66}$$

Solution 5.6

Simplifying, we have

$$\begin{aligned} P_2 &= \begin{bmatrix} \frac{q}{q+p} + \frac{p}{q+p}(1-p-q)^2 & \frac{q}{q+p} - \frac{q}{q+p}(1-p-q)^2 \\ \frac{p}{q+p} - \frac{p}{q+p}(1-p-q)^2 & \frac{p}{q+p} + \frac{q}{q+p}(1-p-q)^2 \end{bmatrix} \\ &= \begin{bmatrix} qp-2p+1+p^2 & -(q-2+p)\,q \\ -(q-2+p)\,p & q^2-2q+qp+1 \end{bmatrix}, \end{aligned}$$

which, in view of the formula in Exercise 5.4, proves that $P_2 = P^2$.

We shall use induction to prove that $P_n = P^n$ for all $n \in \mathbb{N}$. We already know that the assertion true for $n = 1$ (and also for $n = 2$). Suppose that $P_n = P^n$. Then some simple, but tedious algebra gives

$$\begin{aligned} P^{n+1} &= PP^n \\ &= \begin{bmatrix} 1-p & q \\ p & 1-q \end{bmatrix} \begin{bmatrix} \frac{q}{q+p} + \frac{p}{q+p}(1-p-q)^n & \frac{q}{q+p} - \frac{q}{q+p}(1-p-q)^n \\ \frac{p}{q+p} - \frac{p}{q+p}(1-p-q)^n & \frac{p}{q+p} + \frac{q}{q+p}(1-p-q)^n \end{bmatrix} \\ &= \begin{bmatrix} \frac{q}{q+p} + \frac{p}{q+p}(1-p-q)^{n+1} & \frac{q}{q+p} - \frac{q}{q+p}(1-p-q)^{n+1} \\ \frac{p}{q+p} - \frac{p}{q+p}(1-p-q)^{n+1} & \frac{p}{q+p} + \frac{q}{q+p}(1-p-q)^{n+1} \end{bmatrix} \\ &= P_{n+1}. \end{aligned}$$

Solution 5.7

Recall that $P^{n+k} = P^n P^k$. Since $p_{n+k}(j|i)$ are the entries of the matrix $P_{n+k} = P^{n+k} = P^n P^k$, we obtain (5.18) directly from the definition of the product of two matrices.

Solution 5.8

Since the η_i are independent, ξ_n and η_{n+1} are also independent. Therefore,

$$\begin{aligned} & P(\xi_{n+1} = s|\xi_0 = s_0, \cdots, \xi_n = s_n) \\ = \; & P(\xi_n + \eta_{n+1} = s|\xi_0 = s_0, \cdots, \xi_n = s_n) \\ = \; & P(\eta_{n+1} = s - s_n|\xi_0 = s_0, \cdots, \xi_n = s_n) \\ = \; & P(\eta_{n+1} = s - s_n). \end{aligned}$$

Similarly,

$$\begin{aligned} P(\xi_{n+1} = s|\xi_n = s_n) &= P(\xi_n + \eta_{n+1} = s|\xi_n = s_n) \\ &= P(\eta_{n+1} = s - s_n|\xi_n = s_n) \\ &= P(\eta_{n+1} = s - s_n). \end{aligned}$$

Solution 5.9

Step 1. For $n = 1$ the right-hand side of (5.19) is equal to 0 unless $|j - i| \leq 1$ and $1 + j - i$ is even. This is only possible when $j = i + 1$ or $j = i - 1$. In the former case the right-hand side equals p, and in the latter it equals q. This proves (5.19) for $n = 1$.

Step 2. Suppose that (5.19) is true for some n. We will use the following version of the total probability formula. If $H_i \in \mathcal{F}$, $P(H_i \cap H_j) = 0$ for $i \neq j$, and $P(\bigcup_i H_i) = 1$, then

$$P(A|C) = \sum_i P(A|C \cap H_i) P(H_i|C). \tag{5.67}$$

Then the Markov property and (5.67) imply that

$$\begin{aligned} & P(\xi_{n+1} = j|\xi_0 = i) \\ = \; & P(\xi_{n+1} = j|\xi_0 = i, \xi_n = j - 1) P(\xi_n = j - 1|\xi_0 = i) \\ & + P(\xi_{n+1} = j|\xi_0 = i, \xi_n = j + 1) P(\xi_n = j + 1|\xi_0 = i) \\ = \; & P(\xi_{n+1} = j|\xi_n = j - 1) P(\xi_n = j - 1|\xi_0 = i) \\ & + P(\xi_{n+1} = j|\xi_n = j + 1) P(\xi_n = j + 1|\xi_0 = i) \\ = \; & p \binom{n}{\frac{n+j-1-i}{2}} p^{\frac{n+j-1-i}{2}} q^{\frac{n-j+1+i}{2}} + q \binom{n}{\frac{n+j+1-i}{2}} p^{\frac{n+j+1-i}{2}} q^{\frac{n-j-1+i}{2}} \end{aligned}$$

$$= \left[\binom{n}{\frac{n-1+j-i}{2}} + \binom{n}{\frac{n+1+j-i}{2}} \right] p^{\frac{n+1+j-i}{2}} q^{\frac{n+1-j+i}{2}}$$

$$= \binom{n+1}{\frac{n+1+j-i}{2}} p^{\frac{n+1+j-i}{2}} q^{\frac{n+1-j+i}{2}}.$$

Solution 5.10

Denote the right-hand side of (5.24) by $h(x)$. It follows by induction that

$$h^{(k)}(x) = \frac{(2k)!}{k!}(1-4x)^{-1/2-k}, \quad |x| < \frac{1}{4}.$$

On the other hand, h is analytic and

$$h(x) = \sum_{k=0}^{\infty} \frac{1}{k!} h^{(k)}(0) x^k, \quad |x| < \frac{1}{4}.$$

Solution 5.11

We shall translate the problem into the Markov chain language. Denote by S the set of all natural numbers $\mathbb{N} = \{0, 1, 2, \cdots\}$. Let ξ_n denote the number of males in the nth year (or generation), where the present year is called year 0. If $\xi_n = i$, i.e. there are exactly i males in year n, then the probability that there will be j males in the next year is given by

$$P(\xi_{n+1} = j | \xi_n = i) = P(X_1 + \cdots X_i = j), \tag{5.68}$$

where $(X_k)_{k=1}^{\infty}$ is a sequence of independent identically distributed random variables with common distribution

$$P(X_1 = m) = p_m, \quad m \in \mathbb{N}.$$

Hence ξ_n, $n \geq 0$ is a Markov chain on S with transition probabilities

$$p(j|i) = P(X_1 + \cdots X_i = j). \tag{5.69}$$

Notice that $p(0|0) = 1$, i.e. if $\xi_n = 0$, then $\xi_m = 0$ for all $m \geq n$. Dying out means that eventually $\xi_n = 0$, starting from some $n \in \mathbb{N}$. Once this happens, ξ_n will stay at 0 forever.

Solution 5.12

We shall only deal with part 2), as in part 1) there is nothing to show. Suppose that $\xi_0 = i$. Then $\xi_1 = X_1 + \cdots X_i$, where X_j are independent identically distributed Poisson random variables with parameter λ. Since the sum of such random variables has the Poisson distribution with parameter $i\lambda$, (5.27) follows readily.

Solution 5.13

Let ξ_n denote the number of supporters at the end of day n. Then $\xi_{n+1} - \xi_n$ is equal to 1 or 0 and

$$p(j|i) = \begin{cases} pq, & \text{if } j = i+1, \\ 1 - pq, & \text{if } j = i, \\ 0, & \text{otherwise.} \end{cases} \tag{5.70}$$

Thus ξ_n is a Markov chain on a state space $S = \mathbb{N}$ with transition probabilities $p(j|i)$ given by (5.70).

Solution 5.14

We shall use the notation introduced in the hint. Observe that $\xi_{n+1} - \xi_n - Z_n$ equals ± 1 or 0. The latter case occurs with probability p, i.e. when the car served at the beginning of the nth time interval was finished by the end of the nth time interval. The former case occurs with probability $1 - p$. Therefore

$$P(\xi_{n+1} = j|\xi_n = i) = p\frac{\lambda^{j-i+1}}{(j-i+1)!}e^{-\lambda} + (1-p)\frac{\lambda^{j-i}}{(j-i)!}e^{-\lambda}$$

for $j \geq i \geq 1$. On the other hand,

$$P(\xi_{n+1} = j|\xi_n = 0) = \frac{\lambda^j}{j!}e^{-\lambda}$$

if $j \geq i = 0$. Finally, if $j = i - 1$, $i \geq 1$, then

$$P(\xi_{n+1} = i-1|\xi_n = i) = pe^{-\lambda}.$$

Thus, ξ_n is a Markov chain with transition probabilities

$$p(j|i) = \begin{cases} q_j, & \text{if } i = 0, j \in \mathbb{N}, \\ q'_{j-i+1}, & \text{if } i \geq 1, j \in \mathbb{N}, \end{cases} \tag{5.71}$$

where $q_k = \frac{\lambda^k}{k!}e^{-\lambda}$ and

$$q'_k = \begin{cases} pq_k, & \text{if } k = 0, \\ (1-p)q'_{k-1} + pq'_k, & \text{if } k \geq 1. \end{cases} \tag{5.72}$$

The transition probability matrix of our chain takes the form

$$P = \begin{bmatrix} q_0 & q'_0 & 0 & 0 & \cdots & \cdots \\ q_1 & q'_1 & q'_0 & 0 & \cdots & \cdots \\ q_2 & q'_2 & q'_1 & q'_0 & 0 & \cdots \\ \vdots & \vdots & q'_2 & q'_1 & \cdots & \cdots \\ \vdots & \vdots & \vdots & \vdots & \ddots & \\ \vdots & \vdots & \vdots & \vdots & & \ddots \end{bmatrix}.$$

Solution 5.15

If $p_k(j|i) > 0$, then $P(\xi_n = j \text{ for some } n \geq 0|\xi_0 = i) \geq p_k(j|i) > 0$. If $p_k(j|i) = 0$ for all $k \geq 1$, then

$$P(\xi_n = j \text{ for some } n \geq 0|\xi_0 = i) \leq \sum_{n=1}^{\infty} P(\xi_n = j|\xi_0 = i) = \sum_{n=1}^{\infty} p_n(j|i) = 0.$$

Solution 5.16

Since $P(\xi_0 = i|\xi_0 = i) = 1$, it follows that $i \leftrightarrow i$, which proves 1). Assertion 2) is obvious. To prove 3) we proceed as follows. From the solution to Exercise 5.15 we can find $n, m \geq 1$ such that $p_n(j|i) > 0$ and $p_m(k|j) > 0$. Hence, the Chapman–Kolmogorov equations yield

$$p_{m+n}(k|i) = \sum_{s \in S} p_m(k|s) p_n(s|i) \geq p_m(k|j) p_n(j|i) > 0.$$

Solution 5.17

Since $|p_n(j|i)| \leq 1$ and $|f_n(j|i)| \leq 1$ for all $n \in \mathbb{N}$, the radii of convergence of both power series are ≥ 1. To prove the equalities (5.40)–(5.41) we shall show that for $n \geq 1$ and any $i, j \in S$,

$$p_n(j|i) = \sum_{k=1}^{n} f_k(j|i) p_{m-k}(j|j). \tag{5.73}$$

By total probability formula and the Markov property

$$\begin{aligned}
p_n(j|i) &= P(\xi_n = j|\xi_0 = i) \\
&= \sum_{k=1}^{n} P(\xi_n = j, \xi_k = j, \xi_l \neq j, 1 \leq l \leq k-1|\xi_0 = i) \\
&= \sum_{k=1}^{n} P(\xi_k = j, \xi_l \neq j, 1 \leq l \leq k-1|\xi_0 = i) \\
&\qquad \times P(\xi_n = j|\xi_k = j, \xi_l \neq j, 1 \leq l \leq k-1) \\
&= \sum_{k=1}^{n} f_k(j|i) P(\xi_n = j|\xi_k = j) \\
&= \sum_{k=1}^{n} f_k(j|i) p_{n-k}(j|j).
\end{aligned}$$

Solution 5.18

Since $0 \leq p_n(j|i) \leq 1$ and $0 \leq f_n(j|i) \leq 1$, the result follows readily from Abel's lemma.

Solution 5.19

Only the case of a recurrent state needs to be studied. Suppose that j is recurrent. Then $\sum_{n=1}^{\infty} f_n(j|j) = 1$. Hence, by Exercise 5.18, $F_{jj}(x) \nearrow 1$ as $x \nearrow 1$. Thus $P_{jj}(x) = (1 - F_{jj}(x))^{-1} \to \infty$ as $x \nearrow 1$, and so, again by Exercise 5.18, $\sum_{n=1}^{\infty} p_n(j|j) = \infty$. Conversely, suppose that $\sum_{n=1}^{\infty} p_n(j|j) = \infty$. Then, by Exercise 5.18, $P_{jj}(x) = \to \infty$ as $x \nearrow 1$. Thus, $F_{jj}(x) = 1 - (P_{jj}(x))^{-1} \to 1$ as $x \nearrow 1$. Hence, $\sum_{n=1}^{\infty} f_n(j|j) = 1$, which proves that j is recurrent.

To prove (5.43) we use (5.73) to get

$$\begin{aligned}
\sum_{n=0}^{\infty} p_n(1|1) &= \sum_{n=0}^{\infty}\sum_{k=0}^{n-1} f_{n-k}(j|i)p_k(j|j) \\
&= \sum_{k=0}^{\infty}\sum_{m=1}^{\infty} f_m(j|i)p_k(j|j) \\
&= \sum_{k=0}^{\infty} p_k(j|j) \sum_{m=1}^{\infty} f_m(j|i) \\
&\leq \sum_{k=0}^{\infty} p_k(j|j).
\end{aligned}$$

This implies (5.43) when j is transient.

Solution 5.20

We shall show that the state 0 is recurrent. The other case can be treated in a similar way. From Exercise 5.8 we have $p_n(0|0) = \frac{q}{p+q} + \frac{p}{p+q}(1-p-q)^n$. Thus $p_n(0|0) \to \frac{q}{p+q} > 0$, and so $\sum_{n=0}^{\infty} p_n(0|0) = \infty$, which proves that 0 is a recurrent state.

Solution 5.21

Suppose each $j \in S$ is transient. Then by (5.43) $\sum_n p_n(j|i) < \infty$ for all $i \in S$. Let us fix $i \in S$. Then we would have $\sum_{j\in S}\sum_{n=1}^{\infty} p_n(j|i) < \infty$, since S is finite. However, this is impossible because $\sum_{j\in S}\sum_{n=1}^{\infty} p_n(j|i) = \sum_{n=1}^{\infty}\sum_{j\in S} p_n(j|i) = \sum_{n=1}^{\infty} 1 = \infty$.

Solution 5.22

Let us begin with a brief remark concerning the last part of the problem. Since the random walk is 'space homogenous', i.e. $p(j|i) = p(j-i|0)$, it should be quite obvious, at least intuitively, that either all states are positive-recurrent or all states are null-recurrent. One can prove this rigorously without any particular difficulty. First, observe (and prove by induction) that $p_n(j|i) = p_n(j-i|0)$.

Second, observe that the same holds for f_n, i.e. $f_n(j|i) = f_n(j-i|0)$. Hence, in particular, $m_i = m_0$.

To show that 0 is null-recurrent let us recall some useful tools:

$$P_{00}(x) = \sum_{n=0}^{\infty} f_n(0|0)x^n, \quad -1 < x < 1,$$
$$F_{00}(x) = \sum_{n=1}^{\infty} f_n(0|0)x^n, \quad -1 < x < 1.$$

Since, see Exercise 5.9,

$$p_{2k}(0|0) = \binom{2k}{k}\frac{1}{4^k},$$
$$P_{00}(x) = \sum_{k=0}^{\infty} \binom{2k}{k}\left(\frac{x^2}{4}\right)^k = \left(1-x^2\right)^{-1/2}.$$

Then, using (5.41), we infer that $F_{00}(x) = 1-(1-x^2)^{1/2}$. Since $F'_{00}(x) \nearrow \infty$ as $x \nearrow 1$ and $F'_{00}(x) = \sum_{n=1}^{\infty} nf_n(0|0)x^n$ by using Abel's lemma (compare with Exercise 5.18), we infer that $\sum_{n=1}^{\infty} nf_n(0|0) = \infty$. This shows that $m_0 = \infty$ and thus 0 is null-recurrent.

Solution 5.23

We know that 0 is a recurrent state. From definition

$$f_n(0|0) = p(0|1)p(1|1)^{n-2}p(1|0) = pq(1-q)^{n-2}.$$

Since $|1-q| < 1$, we infer that $\sum_{n=1}^{\infty} nf_n(0|0) = \sum_{n=1}^{\infty} pq(1-q)^{n-2} < \infty$. Hence $m_0 < \infty$ and 0 is positive-recurrent. The same proof works for state 1.

Solution 5.24

Consider a Markov chain on $S = \{1, 2, \cdots, 9\}$ with transition probabilities given by the graph in Figure 5.3. Then, obviously, $p_4(1|1) = 1/2$ and $p_6(1|1) = 1/2$,

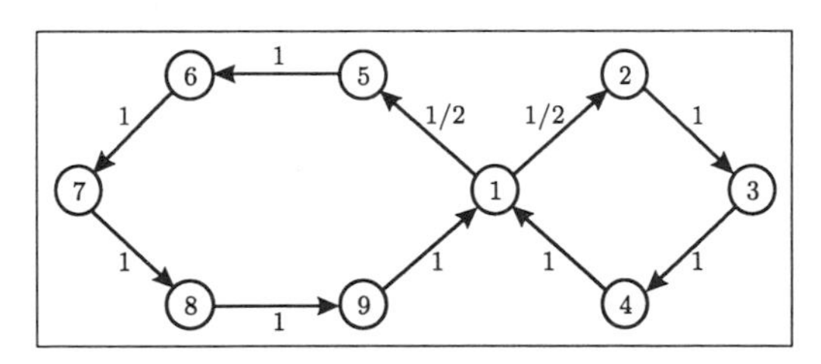

Figure 5.3. Transition probabilities of the Markov chain in Exercise 5.24

but $p_k(1|1) = 0$ if $k \leq 6$ and $k \notin \{4, 6\}$. Hence $d(1) = 2$, but $p_2(1|1) = 0$.

Solution 5.25

We begin with finding $P_2 = P^2$ in an algebraic way, i.e. by multiplying the matrix P by itself. We have

$$P_2 = P^2 = \begin{bmatrix} 0 & 1/2 \\ 1 & 1/2 \end{bmatrix} \begin{bmatrix} 0 & 1/2 \\ 1 & 1/2 \end{bmatrix} = \begin{bmatrix} \frac{1}{2} & \frac{1}{4} \\ \frac{1}{2} & \frac{3}{4} \end{bmatrix}.$$

Alternatively, P_2 can be found by observing that the only way one can get from 1 to 1 in two steps is to move from 1 to 2 (with probability 1) and then from 2 to 1 (with probability 1/2). Hence, the probability $p_2(1|1)$ of going from 1 to 1 in two steps equals 1/2. Analogously, we calculate $p_2(1|2)$ by observing that in order to move from 1 to 2 in two steps one needs first to move from 1 to 2 (with probability 1) and then stay at 2 (with probability 1/2). Hence, $p_2(1|2) = 1/2$. The remaining two elements of the matrix P_2 can be found by repeating the above argument, or, simply by adding the rows so that they equal 1. In the latter method we use the fact that P_2 is a stochastic matrix, see Exercise 5.3. The graph representing P_2 is shown in Figure 5.4.

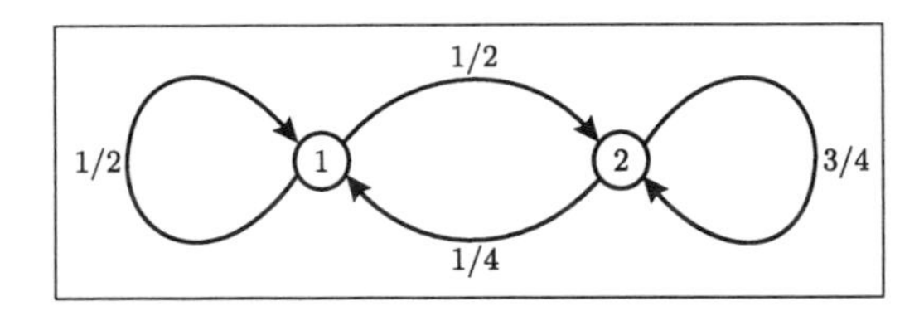

Figure 5.4. Two-step transition probabilities in Exercise 5.25

Using any of the methods presented above, we obtain

$$P_3 = \begin{bmatrix} \frac{1}{4} & \frac{3}{8} \\ \frac{3}{4} & \frac{5}{8} \end{bmatrix}.$$

Therefore, $p_1(1|1) = 0$, $p_2(1|1) = 1/2$ and $p_3(1|1) = 1/4$. Hence $d(1) = 1$ (although $p_1(1|1) = 0$). Since $p_1(2|2) = 1/2 > 0$, it follows that $d(2) = 1$.

Solution 5.26

Suppose that $\xi_0 = i$. Denote by τ_k the minimum positive time when the chain enters state k, i.e.

$$\tau_k = \min\{n \geq 1 : \xi_n = k\}.$$

Then, $P\{\tau_j < \tau_i\} =: \varepsilon > 0$ if $i \to j$ and $i \neq j$. If $j \nrightarrow i$, then it would be impossible to return to i with probability at least $\varepsilon > 0$. But this cannot happen as i is a recurrent state. Indeed,

$$\begin{aligned} 1 &= P(\tau_i < \infty) = P(\tau_i < \infty | \tau_j < \tau_i) P(\tau_j < \tau_i) \\ &\quad + P(\tau_i < \infty | \tau_j \geq \tau_i) P(\tau_j \geq \tau_i). \end{aligned}$$

The second term on the right-hand side is $\leq 1-\varepsilon < 1$, while the first factor in the first term is equal to 0 (since $i \nrightarrow j$). This is a contradiction. The second part is obvious.

Solution 5.27

As observed in the hint, we only need to show that (5.45) holds when $p(j|i) = 0$ for $i \in C$ and $j \in S \setminus C$. We begin with the following simple observation. If $(\Omega, \mathcal{F}, P)$ is a probability space and $A_n \in \mathcal{F}$, $n \in \mathbb{N}$, then $P(\bigcup_n A_n) = 0$ if and only if $P(A_n) = 0$ for each $n \in \mathbb{N}$. Hence (5.45) holds if and only if for each $k \geq n$

$$P\left(\xi_k \in S \setminus C | \xi_n \in C\right) = 0. \tag{5.74}$$

In fact, the above holds if and only if it holds for $k = n+1$. Indeed, suppose that for each $n \in \mathbb{N}$

$$P\left(\xi_{n+1} \in S \setminus C | \xi_n \in C\right) = 0. \tag{5.75}$$

Let us take $n \in \mathbb{N}$. We shall prove by induction on $k \geq n$ that (5.74) holds. This is so for $k = n$ and $k = n+1$. Suppose that (5.74) holds for some $k \geq n+1$. We shall verify that it holds for $k+1$. By the total probability formula (5.67) and the Markov property (5.10)

$$\begin{aligned}
&P\left(\xi_{k+1} \in S \setminus C | \xi_n \in C\right) \\
&\quad = P\left(\xi_{k+1} \in S \setminus C | \xi_n \in C, \xi_k \in C\right) P(\xi_k \in C | \xi_n \in C) \\
&\qquad + P\left(\xi_{k+1} \in S \setminus C | \xi_n \in C, \xi_k \in S \setminus C\right) P(\xi_k \in S \setminus C | \xi_n \in C) \\
&\quad = P\left(\xi_{k+1} \in S \setminus C | \xi_k \in C\right) P(\xi_k \in C | \xi_n \in C) \\
&\qquad + P\left(\xi_{k+1} \in S \setminus C | \xi_k \in S \setminus C\right) P(\xi_k \in S \setminus C | \xi_n \in C).
\end{aligned}$$

By the induction hypothesis $P(\xi_k \in S \setminus C | \xi_n \in C) = 0$ and by (5.75) (applied to k rather than n) $P\left(\xi_{k+1} \in S \setminus C | \xi_k \in C\right) = 0$. Thus, $P\left(\xi_{k+1} \in S \setminus C | \xi_n \in C\right) = 0$, which proves (5.74).

The time-homogeneity of the chain implies that (5.75) is equivalent to (5.74) for $n = 0$. Since P is a countably additive measure and S is a countable set, the latter holds if and only if $p(j|i) = 0$ for $i \in C$ and $j \in S \setminus C$.

Solution 5.28

Property 2) in Proposition 5.8 implies that μ is an invariant measure. Therefore it remains to prove uniqueness. Suppose that $\nu = \sum_{j=1}^{\infty} q_j \delta_j$ is an invariant measure. It is sufficient then to show that $\pi_j = q_j$ for all $j \in S$.
Since $0 \leq p_n(j|i) q_i \leq q_i$ for all $i, j \in S$ and $\sum_i q_i = 1 < \infty$, Lebesgue's dominated convergence theorem yields

$$q_j = \sum_{i=1}^{\infty} p_n(j|i) q_i \to \sum_i \pi_j q_i = \pi_j.$$

It follows that $\mu = \nu$.

Solution 5.29

We shall argue as in the uniqueness part of Solution 5.28. If $\nu = \sum_{j=1}^{\infty} q_j \delta_j$ is an invariant measure, then by Lebesgue's dominated convergence theorem

$$q_j = \sum_{i=1}^{\infty} p_n(j|i) q_i \to \sum_i \pi_j q_i = 0.$$

Hence $\nu = 0$, which contradicts the assumption that ν is a probability measure.

Solution 5.30

Obviously, $P = \begin{bmatrix} 0 & 1 \\ 1 & 0 \end{bmatrix}$. Therefore equation $P\pi = \pi$ becomes

$$\begin{aligned} \pi_1 &= \pi_2, \\ \pi_2 &= \pi_1. \end{aligned}$$

The only solution of this system subject to the condition $\pi_1 + \pi_2 = 1$ is $\pi_1 = \pi_2 = 1/2$.

Solution 5.31

Put $\mu = \sum_{i \in S} \mu_i \delta_i$ and $\nu = \sum_{i \in S} \nu_i \delta_i$. Then, for any $j \in S$ and $n \in \mathbb{N}$

$$\begin{aligned} \sum_{i \in S} p_n(j|i) \left((1-\theta)\mu_i + \theta \nu_i \right) &= (1-\theta) \sum_{i \in S} p_n(j|i) \mu_i + \theta \sum_{i \in S} p_n(j|i) \nu_i \\ &= (1-\theta)\mu_j + \theta \nu_j = \left[(1-\theta)\mu + \theta \nu \right]_j . \end{aligned}$$

Solution 5.32

Put $\mu = \sum_{i \in S} \mu_i \delta_i$. Then, for any $j \in T$

$$\mu_j = \sum_{i \in S} p_n(j|i) \mu_i \to 0$$

by the Lebesgue dominated convergence theorem. Indeed, by Exercise 5.19 $p_n(j|i) \to 0$ for all $i \in S$, and $\sum_{i \in S} p_n(j|i) \mu_i \le \mu_i$, where $\sum \mu_i < \infty$.

Solution 5.33

Since j is transient in view of (5.43), from Exercise 5.19 we readily get (5.53).

Solution 5.34

By Theorem 5.5 $p_n(j|i) \to \frac{1}{m_j}$ for all $i, j \in S$. Put $\pi_j = \frac{1}{m_j}$, $j \in S$. We need to show that

1) $\pi = \sum_j \pi_j \delta_j$ is an invariant measure of the chain ξ_n, $n \in \mathbb{N}$;

2) π is the unique invariant measure.

Part 1) will follow from Proposition 5.8 as soon as we can show that $\pi_j > 0$ for at least one $j \in S$. But if $j \in S$, then j is positive-recurrent and thus $m_j < \infty$. Part 2) follows from Exercise 5.28.

Solution 5.35

Let us fix $i, j \in \mathbb{Z}$. First suppose that $j - i = 2a \in 2\mathbb{Z}$. Then $p_{2k+1}(j|i) = 0$ for all $k \in \mathbb{N}$, and also $p_{2k}(j|i) = 0$ if $k < |a|$. Moreover, if $k \geq |a|$, then

$$p_{2k}(j|i) = \binom{2k}{k+a} p^{k+a} q^{k-a} = \left(\frac{p}{q}\right)^a \binom{2k}{k+a} p^k q^k.$$

Since $\binom{2k}{k+a} \leq \binom{2k}{k}$, it follows that

$$p_{2k}(j|i) \leq \left(\frac{p}{q}\right)^a \binom{2k}{k} p^k q^k \to 0$$

by Proposition 5.4. We have therefore proven that π_j is well defined and that $\pi_j = 0$ for all $j \in \mathbb{Z}$. Hence, we infer that no invariant measure exists.

Solution 5.36

Suppose that $\pi = \sum_{j=0}^{\infty} \pi_j \delta_j$ is an invariant measure of our Markov chain. Then $\sum_i p(j|i)\pi_i = \pi_j$ for all $j \in S = \mathbb{N}$. Using the exact form of the transition probability matrix in the solution to Exercise 5.14, we see that the sequence of non-negative numbers $(\pi_j)_{j=0}^{\infty}$ solves the following infinite system of linear equations:

$$\begin{array}{lcl} q_0\pi_0 + q_0'\pi_1 & = & \pi_0 \\ q_1\pi_0 + q_1'\pi_1 + q_0'\pi_2 & = & \pi_1 \\ q_2\pi_0 + q_2'\pi_1 + q_1'\pi_2 + q_0'\pi_3 & = & \pi_2 \\ \cdots & = & \cdots \end{array} \tag{5.76}$$

i.e.

$$q_k\pi_0 + \sum_{j=1}^{k+1} q_{k+1-j}'\pi_j = \pi_k, \quad k \in \mathbb{N}.$$

Multiplying the kth equation in (5.76) by x^k, $k \geq 0$, and summing all of them up we obtain

$$\pi_0 Q(x) + \frac{G(x)}{x} \sum_{j=1}^{\infty} \pi_j x^j = \Pi(x), \quad |x| < 1, \tag{5.77}$$

where for $|x| < 1$,

$$\begin{aligned} \Pi(x) &= \sum_{j=0}^{\infty} \pi_j x^i, \\ Q(x) &= \sum_{j=0}^{\infty} q_j x^i, \\ G(x) &= \sum_{j=0}^{\infty} q'_j x^i. \end{aligned}$$

Since $\sum_{j=1}^{\infty} \pi_j z^j = \Pi(x) - \pi_0$, we see that Π, Q and G satisfy the functional equation

$$\Pi(x) = \pi_0 \frac{G(x) - xQ(x)}{G(x) - x}, \quad |x| < 1. \tag{5.78}$$

Since all the coefficients π_j in the power series defining Π are non-negative, Abel's lemma implies that

$$\sum_{j=0}^{\infty} \pi_j = \lim_{x \nearrow 1} \Pi(x).$$

It is possible to use (5.78) to calculate this limit. However, there is still some work be done. We have to use l'Hospital's rule. Since $G(x) - x \to 0$, $G(x) - xQ(x) \to 0$, $G'(x) - 1 \to \lambda' - 1$ and $(G(x) - xQ(x))' \to \lambda' - \lambda - 1$ (all limits are for $x \nearrow 1$), we obtain (recall that $\lambda' = \lim_{x \nearrow 1} G'(x)$)

1) if $\lambda' < 1$, then $\Pi(x) \to \frac{1}{1-\lambda'} > 0$;
2) if $\lambda' > 1$, then $\Pi(x) \to \frac{1}{1-\lambda'} < 0$;
3) if $\lambda' = 1$, then $\Pi(x) \to \infty$.

Therefore, we conclude that the invariant measure exists if and only if 1) holds, i.e. $\lambda' < 1$. Then

$$\pi_0 = 1 - \lambda'.$$

If $\lambda' < 1$, then, as we have seen above, there exists a unique invariant measure. Since our chain is irreducible and aperiodic (check this!), it follows from Theorem 5.6 that the Markov chain ξ_n is ergodic. Therefore, by Theorem 5.5, for any $j \in \mathbb{N}$

$$p_n(j|j) \to \pi_j, \quad \text{as } n \to \infty.$$

Since $F_{ji}(1) = 1$, it also follows that

$$p_n(j|i) \to \pi_j, \quad \text{as } n \to \infty.$$

Solution 5.37

We begin with the case $n_0 = 1$. Then $M_n(j) - m_n(j) \le (1-\varepsilon)^n$ from (5.62). It follows that

$$\begin{aligned} m_n(j) &\le \pi_j \le M_n(j), \\ m_n(j) &\le p_n(j|i) \le M_n(j), \end{aligned}$$

and we infer that $|p_n(j|i) - \pi_j| \le (1-\varepsilon)^n$, $n \in \mathbb{N}$. Suppose that $n_0 \ge 2$. Then for any $r = 1, \cdots, n_0 - 1$

$$\begin{aligned} |p_{kn_0+r}(j|i) - \pi_j| &= \left| \sum_{s\in S} p_{kn_0}(j|s) p_r(s|i) - \sum_{s\in S} \pi_j p_r(s|i) \right| \\ &\le \sum_{s\in S} |p_{kn_0}(j|s) - \pi_j| p_r(s|i) \\ &\le \max_{s\in S} |p_{kn_0}(j|s) - \pi_j| \sum_{s\in S} p_r(s|i) \le (1-\varepsilon)^n. \end{aligned}$$

Solution 5.38

The transition probability matrix P takes the form $\begin{bmatrix} 1-p & q \\ p & 1-q \end{bmatrix}$. Since all four numbers $p, q, 1-p, 1-q$ are strictly positive, the assumption (5.54) is satisfied, and so the limits $\pi_j = \lim_n p_n(j|i)$ exist and are i-independent. Hence, the unique invariant measure of the corresponding Markov chain is equal to $\pi_0\delta_0 + \pi_1\delta_1$. (Recall that $S = \{0,1\}$ in this example.) We need to find the values π_j. One way of finding them is to use the definition. As in Hint 5.28, the vector $\pi = (\pi_0, \pi_1)$ solves the following linear equation in matrix form:

$$\begin{bmatrix} 1-p & q \\ p & 1-q \end{bmatrix} \begin{bmatrix} \pi_0 \\ \pi_1 \end{bmatrix} = \begin{bmatrix} \pi_0 \\ \pi_1 \end{bmatrix} \tag{5.79}$$

$$\pi_0 + \pi_1 = 1. \tag{5.80}$$

Some elementary algebra allows us to find the unique solution to the above problem:

$$\pi_0 = \frac{q}{p+q}, \quad \pi_1 = \frac{p}{p+q}. \tag{5.81}$$

Hence, $\frac{q}{p+q}\delta_0 + \frac{p}{p+q}\delta_1$ is the unique invariant measure of the Markov chain.

Solution 5.39

From (5.66) we infer that

$$p_n(0|0) = \frac{q}{q+p} + \frac{p}{q+p}(1-p-q)^n \to \frac{q}{q+p},$$

$$
\begin{aligned}
p_n(1|0) &= \frac{p}{q+p} - \frac{p}{q+p}(1-p-q)^n \to \frac{p}{q+p}, \\
p_n(0|1) &= \frac{q}{q+p} - \frac{q}{q+p}(1-p-q)^n \to \frac{q}{q+p}, \\
p_n(1|1) &= \frac{p}{q+p} + \frac{q}{q+p}(1-p-q)^n \to \frac{p}{q+p}.
\end{aligned}
$$

Hence $\pi_0 = \frac{q}{q+p}$ and $\pi_1 = \frac{p}{q+p}$. This is in agreement with the previous solution.

Solution 5.40

Suppose that $\xi_n = i$. Then the value of ξ_{n+1} depends entirely on the outcome of the next roll of a die, say X_{n+1}, and the value of ξ_n. Since it depends on the past only through the value of ξ_n, intuitively we can see that we are dealing with a Markov chain. To define ξ_n in a precise way consider a sequence of independent identically distributed random variables X_n, $n = 1, 2, 3, \cdots$ such that $P(X_1 = i) = 1/6$ for all $i = 1, 2, \cdots, 6$. Then, putting $\xi_{n+1} = \max\{\xi_n, X_{n+1}\}$, we can see immediately that $P(\xi_{n+1} = i_{n+1}|\xi_0 = i_0, \cdots, \xi_n = i_n) = P(\xi_{n+1} = i_{n+1}|\xi_n = i_n)$ and

$$
\begin{aligned}
P(\xi_{n+1} = j|\xi_n = i) &= P(\xi_{n+1} = j|\xi_n = i, X_{n+1} \le i)P(X_{n+1} \le i) \\
&\quad + P(\xi_{n+1} = j|\xi_n = i, X_{n+1} > i)P(X_{n+1} > i) \\
&= \frac{i}{6}\begin{cases} 0 & \text{if } j > i, \\ 1 & \text{if } j = i, \\ 0 & \text{if } j < i, \end{cases} + \frac{6-i}{6}\begin{cases} \frac{1}{6-i} & \text{if } j > i, \\ 0 & \text{if } j \le i, \end{cases} \\
&= \begin{cases} \frac{1}{6} & \text{if } j > i, \\ \frac{i}{6} & \text{if } j = i, \\ 0 & \text{if } j < i. \end{cases}
\end{aligned}
$$

Thus,

$$
p(j|i) = \begin{cases} \frac{1}{6} & \text{if } j > i, \\ \frac{i}{6} & \text{if } j = i, \\ 0 & \text{if } j < i. \end{cases}
$$

It follows that

$$
P = \begin{bmatrix}
\frac{1}{6} & 0 & 0 & 0 & 0 & 0 \\
\frac{1}{6} & \frac{2}{6} & 0 & 0 & 0 & 0 \\
\frac{1}{6} & \frac{1}{6} & \frac{3}{6} & 0 & 0 & 0 \\
\frac{1}{6} & \frac{1}{6} & \frac{1}{6} & \frac{4}{6} & 0 & 0 \\
\frac{1}{6} & \frac{1}{6} & \frac{1}{6} & \frac{1}{6} & \frac{5}{6} & 0 \\
\frac{1}{6} & \frac{1}{6} & \frac{1}{6} & \frac{1}{6} & \frac{1}{6} & 1
\end{bmatrix}.
$$

Solution 5.41

The graph of the chain is given in Figure 5.5. The transition matrix is

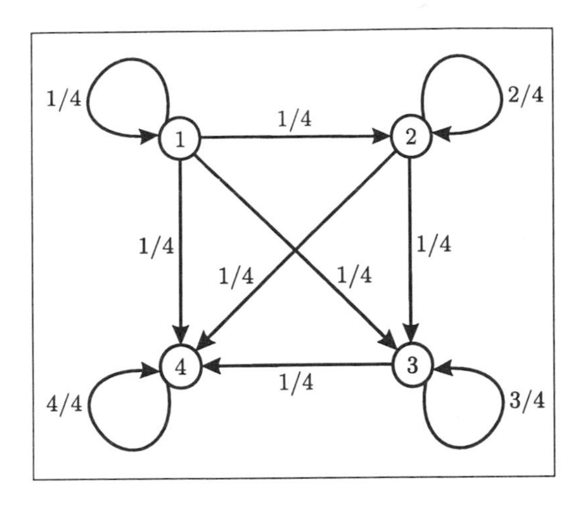

Figure 5.5. Transition probabilities of the Markov chain in Exercise 5.41

$$P = \begin{bmatrix} \frac{1}{4} & 0 & 0 & 0 \\ \frac{1}{4} & \frac{2}{4} & 0 & 0 \\ \frac{1}{4} & \frac{1}{4} & \frac{3}{4} & 0 \\ \frac{1}{4} & \frac{1}{4} & \frac{1}{4} & 1 \end{bmatrix}$$

Either from the graph or from the matrix we can see that $i \to j$ if and only if $i \leq j$. The state $i = 1$ is transient. Indeed, $p(1|1) = 1/4$ and $p_n(1|1) = (1/4)^n$ by induction. Therefore, $\sum_n p_n(1|1) < \infty$ and 1 is transient by Exercise 5.19. The same argument shows that the states 2 and 3 are also transient. On the other hand, $p(4|4) = 1$, and so 4 is a positive-recurrent state. (As we know, that there should be at least one positive-recurrent state.)

We shall find invariant measures by solving the system of four linear equations $P\pi = \pi$ for $\pi = (\pi_1, \pi_2, \pi_3, \pi_4)$, subject to the condition $\pi_1 + \pi_2 + \pi_3 + \pi_4 = 1$. Some elementary linear algebra shows that the only solution is $\pi_1 = \pi_2 = \pi_3 = 0$, $\pi_4 = 1$. Thus, the unique invariant measure is $\pi = \delta_4$.

The invariant measure can also be found by invoking Theorem 5.4. In our case $C = \{4\}$, and so there is exactly one class of recurrent sets. Moreover, as we have seen before, 4 is positive-recurrent. Therefore, there exists a unique invariant measure. Since its support is contained in C, we infer that $\pi = \delta_4$.

Solution 5.42

The graph representing the Markov chain for $a = 4$ is presented in Figure 5.6. Obviously, $i \to j$ for $i \in S \setminus \{0, a\} =: \overset{\circ}{S}$ and $j \in S$. Moreover, $0 \to j$ if and only if $j = 0$ and $a \to j$ if and only if $j = a$. Since $p(0|0) = p(a|a) = 1$, both states 0 and a are positive-recurrent. All other states are transient. For if $i \in \overset{\circ}{S}$ were recurrent, i would be intercommunicating with 0 because $i \to 0$, by Exercise 5.26. This is impossible. Therefore, by Remark 5.5 there exist an

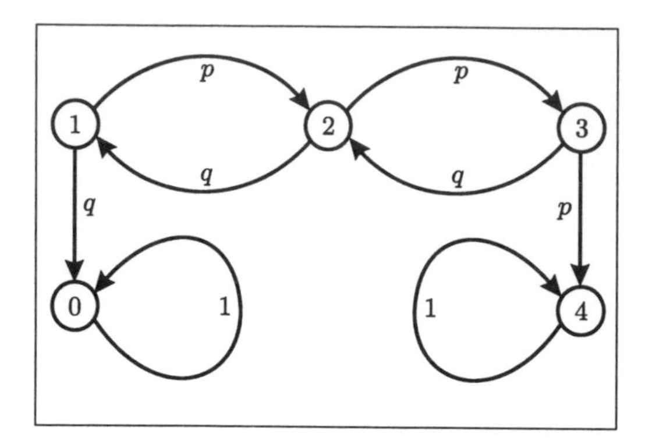

Figure 5.6. Transition probabilities of the Markov chain in Exercise 5.42

infinite number of invariant measures: $\mu = (1-\theta)\delta_0 + \theta\delta_a$, where $\theta \in [0,1]$. Indeed, δ_a is the only probability measure with support in the singleton $\{a\}$. It is possible to verify this with bare hands. Next, let $\phi(i)$ denote the probability that the investigated Markov chain ξ_n hits the right-hand barrier prior to hitting the left-hand one. Once ξ_n hits the left barrier it will never leave it, so $\phi(i)$ is actually the probability that ξ_n hits a. Put

$$A = \{\exists n \in \mathbb{N} : \xi_n = a\}.$$

Then, by the total probability formula and the Markov property of ξ_n, we have for $1 \le i \le a-1$

$$\begin{aligned} \phi(i) &= P(A|\xi_0 = i) = \sum_{j=0}^{a} P(A|\xi_0 = i, \xi_1 = j)P(\xi_1 = j|\xi_0 = i) \\ &= \sum_{j=0}^{a} P(A|\xi_1 = j)P(\xi_1 = j|\xi_0 = i) \\ &= \sum_{j=0}^{a} \phi(j)P(\xi_1 = j|\xi_0 = i) = p\phi(i+1) + q\phi(i-1). \end{aligned}$$

Obviously,

$$\begin{aligned} \phi(0) &= 0, \\ \phi(a) &= 1. \end{aligned}$$

Therefore, the sequence $(\phi(i))_{i=0}^{a}$ satisfies

$$\begin{aligned} \phi(i) &= p\phi(i+1) + q\phi(i-1), \ \ 1 \le i \le a-1, \\ \phi(0) &= 0, \ \phi(a) = 1. \end{aligned} \tag{5.82}$$

Since $p+q=1$, equations (5.82) can be rewritten as follows:

$$p\left[\phi(i+1) - \phi(i)\right] = q\left[\phi(i) - \phi(i-1)\right], \ \ 1 \le i \le a-1.$$

Hence,

$$[\phi(i+1) - \phi(i)] = \left(\frac{q}{p}\right) ix,$$

where $x = \phi(i) - \phi(0)$ is to be determined. Using the boundary condition $\phi(a) = 1$, we can easily find that

$$\begin{aligned} 1 = \phi(a) &= \sum_{i=0}^{a-1} (\phi(i+1) - \phi(i)) \\ &= \frac{(\frac{q}{p})^a - 1}{\frac{q}{p} - 1} x. \end{aligned}$$

Here we assume that $p \neq q$. The case $p = q = 1/2$ can be treated in a similar way. In fact the latter is easier. It follows that

$$x = \frac{\frac{q}{p} - 1}{(\frac{q}{p})^a - 1}$$

and therefore

$$\phi(i) = \sum_{k=0}^{i-1} (\phi(k+1) - \phi(k)) \frac{(\frac{q}{p})^i - 1}{(\frac{q}{p})^a - 1}.$$

6

Stochastic Processes in Continuous Time

6.1 General Notions

The following definitions are straightforward extensions of those introduced earlier for sequences of random variables, the underlying idea being that of a family of random variables depending on time.

Definition 6.1

A *stochastic process* is a family of random variables $\xi(t)$ parametrized by $t \in T$, where $T \subset \mathbb{R}$. When $T = \{1, 2, \ldots\}$, we shall say that $\xi(t)$ is a stochastic process in *discrete time* (i.e. a sequence of random variables). When T is an interval in $\mathbb{R}$ (typically $T = [0, \infty)$), we shall say that $\xi(t)$ is a stochastic process in *continuous time.*

For every $\omega \in \Omega$ the function

$$T \ni t \mapsto \xi(t, \omega)$$

is called a *path* (or *sample path*) of $\xi(t)$.

Definition 6.2

A family $\mathcal{F}_t$ of σ-fields on Ω parametrized by $t \in T$, where $T \subset \mathbb{R}$, is called a *filtration* if

$$\mathcal{F}_s \subset \mathcal{F}_t \subset \mathcal{F}$$

for any $s, t \in T$ such that $s \leq t$.

Definition 6.3

A stochastic process $\xi(t)$ parametrized by $t \in T$ is called a *martingale* (*submartingale*, *supermartingale*) with respect to a filtration $\mathcal{F}_t$ if

1) $\xi(t)$ is integrable for each $t \in T$;
2) $\xi(t)$ is $\mathcal{F}_t$-measurable for each $t \in T$ (in which case we say that $\xi(t)$ is *adapted* to $\mathcal{F}_t$);
3) $\xi(s) = E(\xi(t)|\mathcal{F}_s)$ (respectively, $\leq$ or $\geq$) for every $s, t \in T$ such that $s \leq t$.

In earlier chapters we have seen various stochastic processes, in particular, martingales in discrete time such as the symmetric random walk, for example. In what follows we shall study in some detail two processes in continuous time, namely, the Poisson process and Brownian motion.

6.2 Poisson Process

6.2.1 Exponential Distribution and Lack of Memory

Definition 6.4

We say that a random variable η has the *exponential distribution* of *rate* $\lambda > 0$ if

$$P\{\eta > t\} = e^{-\lambda t}$$

for all $t \geq 0$.

For example, the emissions of particles by a sample of radioactive material (or calls made at a telephone exchange) occur at random times. The probability that no particle is emitted (no call is made) up to time t is known to decay exponentially as t increases. That is to say, the time η of the first emission has the exponential distribution, $P\{\eta > t\} = e^{-\lambda t}$.

Exercise 6.1

What is the distribution function of a random variable η with exponential distribution? Does it have a density? If so, find the density.

Hint What is the probability that $\eta > 0$? What is the probability that $\eta > t$ for any given $t < 0$? Can you express the distribution function in terms of $P\{\eta > t\}$? Is the distribution function differentiable?

Exercise 6.2

Compute the expectation and variance of a random variable having the exponential distribution.

Hint Use the density found in Exercise 6.1.

Exercise 6.3

Show that a random variable η with exponential distribution satisfies

$$P\{\eta > t + s\} = P\{\eta > t\} P\{\eta > s\} \tag{6.1}$$

for any $s, t \geq 0$.

Hint When the probabilities are replaced by exponents, the equality should become obvious.

Exercise 6.4

Show that the equality in Exercise 6.3 is equivalent to

$$P\{\eta > t + s | \eta > s\} = P\{\eta > t\} \tag{6.2}$$

for any $s, t \geq 0$.

Hint Recall how to compute conditional probability. Observe that $\eta > s + t$ implies $\eta > s$.

The equality (6.2) (or, equivalently, (6.1)) is known as the *lack of memory property*. The odds that no particle will be emitted (no call will be made) in the next time interval of length t are not affected by the length of time s it has already taken to wait, given that no emission (no call) has occurred yet.

Exercise 6.5

Show that the exponential distribution is the only probability distribution satisfying the lack of memory property.

Hint The lack of memory property means that $g(t) = P\{\eta > t\}$ satisfies the functional equation

$$g(t + s) = g(t)g(s)$$

for any $s, t > 0$. Find all non-negative non-increasing solutions of this functional equation.

6.2.2 Construction of the Poisson Process

Let $\eta_1, \eta_2, \ldots$ be a sequence of independent random variables, each having the same exponential distribution of rate λ. For example, the times between the emissions of radioactive particles (or between calls made at a telephone exchange) have this property. We put

$$\xi_n = \eta_1 + \cdots + \eta_n,$$

which can be thought of as the time of the nth emission (the nth call). We also put $\xi_0 = 0$ for convenience. The number of emissions (calls) up to time $t \geq 0$ is an n such that $\xi_{n+1} > t \geq \xi_n$. In other words, the number of emissions (calls) up to time $t \geq 0$ is equal to $\max\{n : t \geq \xi_n\}$.

Definition 6.5

We say that $N(t)$, where $t \geq 0$, is a *Poisson process* if

$$N(t) = \max\{n : t \geq \xi_n\}.$$

Thus, $N(t)$ can be regarded as the number of particles emitted (calls made) up to time t. It is an example of a stochastic process in continuous time. A typical path of $N(t)$ is shown in Figure 6.1. It begins at the origin, $N(0) = 0$ (no particles emitted at time 0), and is right-continuous, non-decreasing and piecewise constant with jumps of size 1 at the times ξ_n.

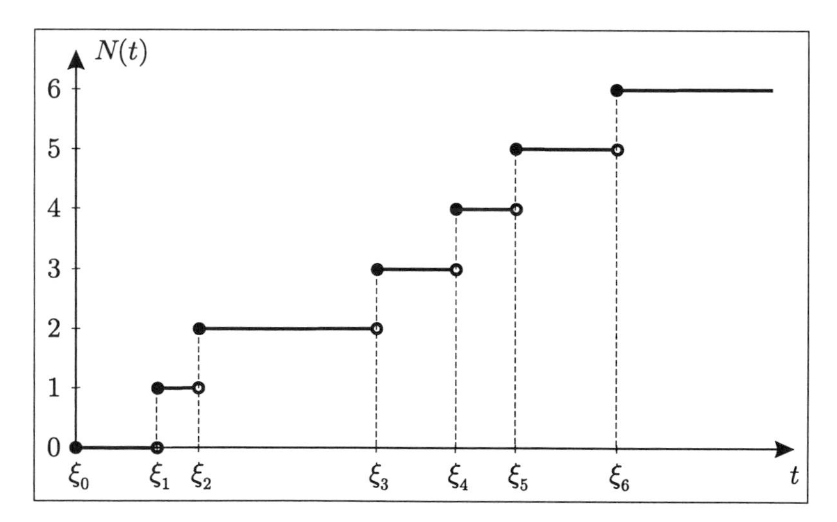

Figure 6.1. A typical path of $N(t)$ and the jump times ξ_n

What is the distribution of $N(t)$? To answer this question we need to recall the definition of the Poisson distribution.

Definition 6.6

A random variable ν has the *Poisson distribution* with parameter $\alpha > 0$ if

$$P\{\nu = n\} = e^{-\alpha}\frac{\alpha^n}{n!}$$

for any $n = 0, 1, 2, \ldots$.

The probabilities $P\{\nu = n\}$ for various values of α are shown in Figure 6.2.

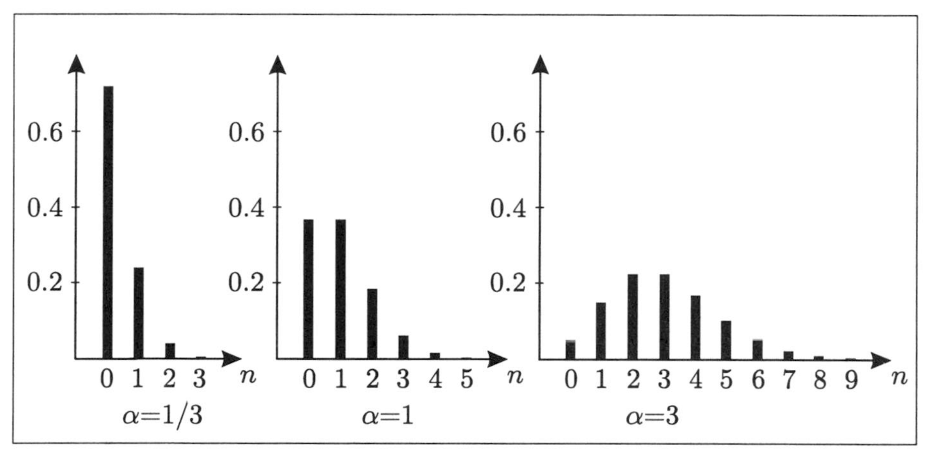

Figure 6.2. Poisson distribution with parameter α

Proposition 6.1

$N(t)$ has the Poisson distribution with parameter λt,

$$P\{N(t) = n\} = e^{-\lambda t}\frac{(\lambda t)^n}{n!}.$$

Proof

First of all, observe that

$$\{N(t) < n\} = \{\xi_n > t\}.$$

It suffices to compute the probability of this event for any n because

$$\begin{aligned} P\{N(t)=n\} &= P\{N(t)<n+1\}-P\{N(t)<n\} \\ &= P\{\xi_{n+1}>t\}-P\{\xi_n>t\}. \end{aligned} \tag{6.3}$$

We shall prove by induction on n that

$$P\{\xi_n>t\}=e^{-\lambda t}\sum_{k=0}^{n-1}\frac{(\lambda t)^k}{k!}. \tag{6.4}$$

For $n=1$

$$P\{\xi_1>t\}=P\{\eta_1>t\}=e^{-\lambda t}.$$

Next, suppose that (6.4) holds for some n. Then, expressing ξ_{n+1} as the sum of the independent random variables ξ_n and η_{n+1}, we compute

$$\begin{aligned} P\{\xi_{n+1}>t\} &= P\{\xi_n+\eta_{n+1}>t\} \\ &= P\{\eta_{n+1}>t\}+P\{\xi_n>t-\eta_{n+1},t\geq\eta_{n+1}>0\} \\ &= e^{-\lambda t}+\int_0^t P\{\xi_n>t-s\}\,f_{\eta_{n+1}}(s)\,ds \\ &= e^{-\lambda t}+\int_0^t e^{-\lambda(t-s)}\sum_{k=0}^{n-1}\frac{(\lambda(t-s))^k}{k!}\lambda e^{-\lambda s}ds \\ &= e^{-\lambda t}+e^{-\lambda t}\sum_{k=0}^{n-1}\frac{\lambda^{k+1}}{k!}\int_0^t (t-s)^k\,ds \\ &= e^{-\lambda t}\sum_{k=0}^{n}\frac{(\lambda t)^k}{k!}, \end{aligned}$$

where $f_{\eta_{n+1}}(s)$ is the density of η_{n+1}. By induction, (6.4) holds for any n. Now apply (6.3) to complete the proof. □

Exercise 6.6

What is the expectation of $N(t)$?

Hint What are the possible values of $N(t)$? What are the corresponding probabilities? Can you compute the expectation from these? To simplify the result use the Taylor expansion of e^x.

Exercise 6.7

Compute $P\{N(s)=1,N(t)=2\}$ for any $0\leq s<t$.

Hint Express $\{N(s)=1,N(t)=2\}$ as $\{\eta_1\leq s<\eta_1+\eta_2\leq t<\eta_1+\eta_2+\eta_3\}$. You can compute the probability of the latter, since η_1,η_2,η_3 are exponentially distributed and independent.

6.2.3 Poisson Process Starts from Scratch at Time t

Imagine that you are to take part in an experiment to count the emissions of a radioactive particle. Unfortunately, in the excitement of proving the lack of memory property you forget about the engagement and arrive late to find that the experiment has already been running for time t and you have missed the first $N(t)$ emissions. Determined to make the best of it, you start counting right away, so at time $t+s$ you will have registered $N(t+s)-N(t)$ emissions. It will now be necessary to discuss $N(t+s)-N(t)$ instead of $N(s)$ in your report.

What are the properties of $N(t+s)-N(t)$? Perhaps you can guess something from the physical picture? After all, a sample of radioactive material will keep emitting particles no matter whether anyone cares to count them or not. So the moment when someone starts counting does not seem important. You can expect $N(t+s)-N(t)$ to behave in a similar way as $N(s)$. And because radioactive emissions have no memory of the past, $N(t+s)-N(t)$ should be independent of $N(t)$.

To study this conjecture recall the construction of a Poisson process $N(t)$ based on a sequence of independent random variables $\eta_1, \eta_2, \ldots$, all having the same exponential distribution. We shall try to represent $N(t+s)-N(t)$ in a similar way.

Let us put

$$\eta_1^t := \xi_{N(t)+1} - t, \quad \eta_n^t := \eta_{N(t)+n}, \quad n = 2, 3, \ldots,$$

see Figure 6.3. These are the times between the jumps of $N(t+s)-N(t)$. Then we define

$$\begin{aligned} \xi_n^t &= \eta_1^t + \cdots + \eta_n^t, \\ N^t(s) &= \max\{n : \xi_n^t \le s\}. \end{aligned}$$

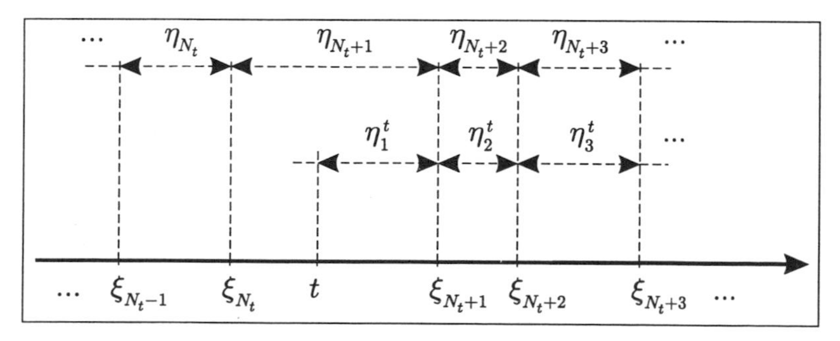

Figure 6.3. The random variables $\eta_1^t, \eta_2^t, \eta_3^t, \ldots$

Exercise 6.8

Show that

$$N^t(s) = N(t+s) - N(t).$$

Hint First show that $\xi_n^t = \xi_{N(t)+n} - t$.

If we can show that $\eta_1^t, \eta_2^t, \ldots$ are independent random variables having the same exponential distribution as $\eta_1, \eta_2, \ldots$, it will mean that $N(t+s) - N(t)$ is a Poisson process with the same probability distribution as $N(s)$. Moreover, if the η_n^t turn out to be independent of $N(t)$, it will imply that $N(t+s) - N(t)$ is also independent of $N(t)$.

Before setting about this task beware of one common mistake. It is sometimes claimed that the times $\eta_1^t, \eta_2^t, \eta_3^t, \ldots$ between the jumps of $N(t+s) - N(t)$ are equal to $\xi_{n+1} - t, \eta_{n+2}, \eta_{n+3}, \ldots$ for some n. Hence $\eta_1^t, \eta_2^t, \eta_3^t, \ldots$ are independent because the random variables $\xi_{n+1}, \eta_{n+2}, \eta_{n+3}, \ldots$ are. The flaw in this is that, in fact, $\eta_1^t, \eta_2^t, \eta_3^t, \ldots$ are equal to $\xi_{n+1} - t, \eta_{n+2}, \eta_{n+3}, \ldots$ only on the set $\{N(t) = n\}$. However, the argument can be saved by conditioning with respect to $N(t)$. Our task becomes an exercise in computing conditional probabilities.

Exercise 6.9

Show that

$$P\{\eta_1^t > s | N(t)\} = P\{\eta_1 > s\}.$$

Hint It suffices (why?) to verify that

$$P\{\eta_1^t > s, N(t) = n\} = P\{\eta_1 > s\} P\{N(t) = n\}$$

for any n. To this end, write the sets $\{N(t) = n\}$ and $\{\eta_1^t > s, N(t) = n\}$ in terms of ξ_n and η_{n+1}, which are independent random variables, and use the lack of memory property for η_{n+1}.

Exercise 6.10

Show that

$$P\{\eta_1^t > s_1, \ldots, \eta_k^t > s_k | N(t)\} = P\{\eta_1 > s_1\} \cdots P\{\eta_k > s_k\}.$$

Hint Verify that

$$P\{\eta_1^t > s_1, \eta_2^t > s_2, \ldots, \eta_k^t > s_k, N(t) = n\} = P\{\eta_1 > s_1\} \cdots P\{\eta_k > s_k\} P\{N(t) = n\}$$

for any n. This is done in Exercise 6.9 for $k = 1$.

Exercise 6.11

From the formula in Exercise 6.10 deduce that the random variables η_n^t and $N(t)$ are independent, and the η_n^t have the same probability distribution as η_n, i.e. exponential with parameter λ.

Hint What do you get if you take the expectation on both sides of the equality in Exercise 6.10? Can you deduce the probability distribution of η_n^t? Can you see that the η_n^t are independent?

To prove that the η_n^t are independent of $N(t)$ you need to be a little more careful and integrate over $\{N(t) = n\}$ instead of taking the expectation.

Because $N(t+s) - N(t)$ can be defined in terms of $\eta_1^t, \eta_2^t, \ldots$ in the same way the original Poisson process $N(t)$ is defined in terms of $\eta_1, \eta_2, \ldots$, the result in Exercise 6.11 proves the theorem below.

Theorem 6.1

For any fixed $t \geq 0$

$$N^t(s) = N(t+s) - N(t), \quad s \geq 0$$

is a Poisson process independent of $N(t)$ with the same probability law as $N(s)$.

That is to say, for any $s, t \geq 0$ the increment $N(t+s) - N(t)$ is independent of $N(t)$ and has the same probability distribution as $N(s)$. The assertion can be generalized to several increments, resulting in the following important theorem.

Theorem 6.2

For any $0 \leq t_1 \leq t_2 \leq \cdots \leq t_n$ the increments

$$N(t_1), N(t_2) - N(t_1), N(t_3) - N(t_2), \ldots, N(t_n) - N(t_{n-1})$$

are independent and have the same probability distribution as

$$N(t_1), N(t_2 - t_1), N(t_3 - t_2), \ldots, N(t_n - t_{n-1}).$$

Proof

From Theorem 6.1 it follows immediately that each increment $N(t_i) - N(t_{i-1})$ has the same distribution as $N(t_i - t_{i-1})$ for $i = 1, \ldots, n$.

It remains to prove independence. This can be done by induction. The case when $n = 2$ is covered by Theorem 6.1. Now suppose that independence holds for n increments of a Poisson process for some $n \geq 2$. Take any sequence

$$0 \leq t_1 \leq t_2 \leq \cdots \leq t_n \leq t_{n+1}.$$

By the induction hypothesis

$$N(t_{n+1}) - N(t_n), \ldots, N(t_2) - N(t_1)$$

are independent, since they can be regarded as increments of

$$N^{t_1}(s) = N(t_1 + s) - N(t_1),$$

which is a Poisson process by Theorem 6.1. By the same theorem these increments are independent of $N(t_1)$. It follows that the $n+1$ random variables

$$N(t_{n+1}) - N(t_n), \ldots, N(t_2) - N(t_1), N(t_1)$$

are independent, completing the proof. □

Definition 6.7

We say that a stochastic process $\xi(t)$, where $t \in T$, has *independent increments* if

$$\xi(t_1) - \xi(t_0), \ldots, \xi(t_n) - \xi(t_{n-1})$$

are independent for any $t_0 < t_1 < \cdots < t_n$ such that $t_0, t_1, \ldots, t_n \in T$.

Definition 6.8

A stochastic process $\xi(t)$, where $t \in T$, is said to have *stationary increments* if for any $s, t \in T$ the probability distribution of $\xi(t+h) - \xi(s+h)$ is the same for each h such that $s+h, t+h \in T$.

Theorem 6.2 implies that the Poisson process has stationary independent increments. The result in the next exercise is also a consequence of Theorem 6.2.

Exercise 6.12

Show that $N(t) - \lambda t$ is a martingale with respect to the filtration $\mathcal{F}_t$ generated by the family of random variables $\{N(s) : s \in [0, t]\}$.

Hint Observe that $N(t) - N(s)$ is independent of $\mathcal{F}_s$ by Theorem 6.2.

6.2.4 Various Exercises on the Poisson Process

Exercise 6.13

Show that $\xi_0 < \xi_1 < \xi_2 < \cdots$ a.s.

Hint What is the probability of the event $\{\xi_{n-1} < \xi_n\} = \{\eta_n > 0\}$? What is the probability of the intersection of all such events?

Exercise 6.14

Show that $\lim_{n\to\infty} \xi_n = \infty$ a.s.

Hint If $\lim_{n\to\infty} \xi_n < \infty$, then the sequence $\eta_1, \eta_2, \ldots$ of independent random variables, all having the same exponential distribution, must be bounded. What is the probability that such a sequence is bounded? Begin with computing the probability $P\{\eta_1 \leq m, \ldots, \eta_n \leq m\}$ for any fixed $m > 0$.

Although it is instructive to estimate $P\{\lim_{n\to\infty} \xi_n < \infty\}$ in this way, there is a more elegant argument based on the strong law of large numbers. What does the law of large numbers tell us about the limit of $\frac{\xi_n}{n}$ as $n \to \infty$?

Exercise 6.15

Show that ξ_n has absolutely continuous distribution with density

$$f_n(t) = \lambda e^{-\lambda t} \frac{(\lambda t)^{n-1}}{(n-1)!}$$

with parameters n and λ. The density $f_n(t)$ of the gamma distribution is shown in Figure 6.4 for $n = 2, 4$ and $\lambda = 1$.

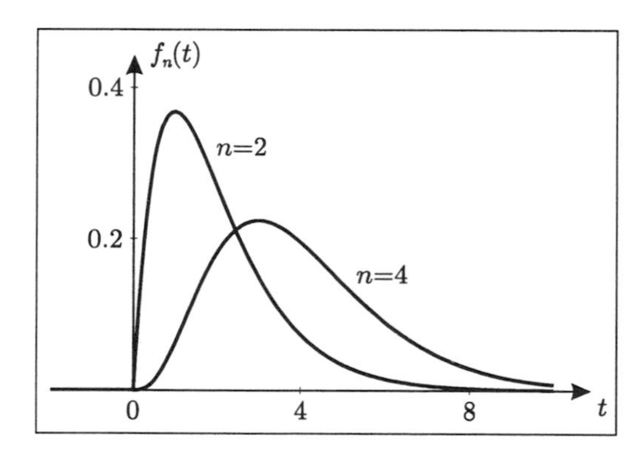

Figure 6.4. Density $f_n(t)$ of the gamma distribution with parameters $n = 2, \lambda = 1$ and $n = 4, \lambda = 1$

Hint Use the formula for $P\{\xi_n > t\}$ in the proof of Proposition 6.1 to find the distribution function of ξ_n. Is this function differentiable? What is the derivative?

Exercise 6.16

Prove that

$$\lim_{t\to\infty} N(t) = \infty \quad \text{a.s.}$$

Hint What is the limit of $P\{N(k) \geq n\}$ as $k \to \infty$? Can you express $\{\lim_{t\to\infty} N(t) = \infty\}$ in terms of the sets $\{N(k) \geq n\}$?

Exercise 6.17

Verify that

$$\begin{aligned} P\{N(t) \text{ is odd}\} &= e^{-\lambda t}\sinh(\lambda t), \\ P\{N(t) \text{ is even}\} &= e^{-\lambda t}\cosh(\lambda t). \end{aligned}$$

Hint What is the probability that $N(t) = 2n+1$? Compare this to the n-th term of the Taylor expansion of $\sinh \lambda t$.

Exercise 6.18

Show that

$$\lim_{t\to\infty} \frac{N(t)}{t} = \lambda \quad \text{a.s.}$$

if $N(t)$ is a Poisson process with parameter λ.

Hint $N(n)$ is the sum of independent identically distributed random variables $N(1)$, $N(2) - N(1), \ldots, N(n) - N(n-1)$, so the strong law of large numbers can be applied to obtain the limit of $N(n)/n$ as $n \to \infty$. Because $N(t)$ is non-decreasing, the limit will not be affected if n is replaced by a continuous parameter $t \geq 0$.

6.3 Brownian Motion

Imagine a cloud of smoke in completely still air. In time, the cloud will spread over a large volume, the concentration of smoke varying in a smooth manner. However, if a single smoke particle is observed, its path turns out to be extremely rough due to frequent collisions with other particles. This exemplifies two aspects of the same phenomenon called diffusion: erratic particle trajectories at the microscopic level, giving rise to a very smooth behaviour of the density of the whole ensemble of particles. The Wiener process $W(t)$ defined below is a mathematical device designed as a model of the motion of individual diffusing particles. In particular, its paths exhibit similar erratic behaviour to the trajectories of real smoke particles. Meanwhile, the density $f_{W(t)}$ of the random variable $W(t)$ is very smooth, given by the exponential function

$$f_{W(t)}(x) = \frac{1}{\sqrt{2\pi t}} e^{-\frac{x^2}{2t}},$$

which is a solution of the *diffusion equation*

$$\frac{\partial f}{\partial t} = \frac{1}{2}\frac{\partial^2 f}{\partial x^2}$$

and can be interpreted as the density at time t of a cloud of smoke issuing form single point source at time 0. The Wiener process $W(t)$ is also associated with the name of the British botanist Robert Brown, who around 1827 observed the random movement of pollen particles in water. We shall study mainly the one-dimensional Wiener process, which can be thought of as the projection of the position of a smoke particle onto one of the axes of a coordinate system.

Apart from describing the motion of diffusing particles, the Wiener process is widely applied in mathematical models involving various noisy systems, for example, the behaviour of asset prices at the stock exchange. If the noise in the system is due to a multitude of independent random changes, then the Central Limit Theorem predicts that the net result will have the normal distribution, a property shared by the increments $W(t) - W(s)$ of the Wiener process. This is one of the main reasons of the widespread use of $W(t)$ in mathematical models.

6.3.1 Definition and Basic Properties

Definition 6.9

The *Wiener process* (or *Brownian motion*) is a stochastic process $W(t)$ with values in $\mathbb{R}$ defined for $t \in [0, \infty)$ such that

1) $W(0) = 0$ a.s.;

2) the sample paths $t \mapsto W(t)$ are a.s. continuous;

3) for any finite sequence of times $0 < t_1 < \cdots < t_n$ and Borel sets $A_1, \ldots, A_n \subset \mathbb{R}$

$$\begin{aligned} &P\{W(t_1) \in A_1, \ldots, W(t_n) \in A_n\} \\ &= \int_{A_1} \cdots \int_{A_n} p(t_1, 0, x_1)\, p(t_2 - t_1, x_1, x_2) \cdots \\ &\qquad \cdots p(t_n - t_{n-1}, x_{n-1}, x_n)\, dx_1 \cdots dx_n, \end{aligned}$$

where

$$p(t, x, y) = \frac{1}{\sqrt{2\pi t}} e^{-\frac{(x-y)^2}{2t}} \tag{6.5}$$

defined for any $x, y \in \mathbb{R}$ and $t > 0$ is called the *transition density*.

A typical sample path of the Wiener process is shown in Figure 6.5.

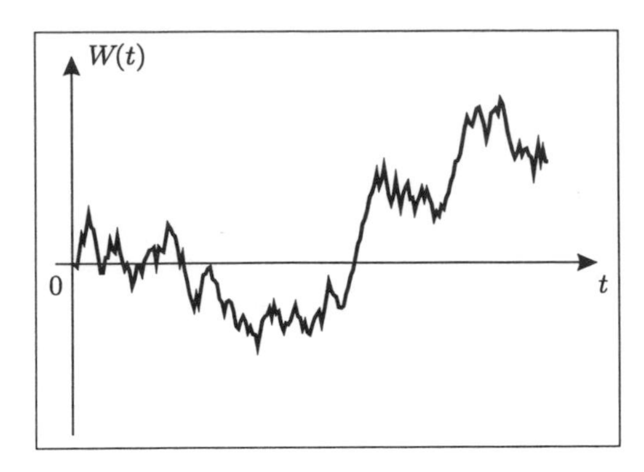

Figure 6.5. A typical path of $W(t)$

Exercise 6.19

Show that

$$f_{W(t)}(x) = \frac{1}{\sqrt{2\pi t}} e^{-\frac{x^2}{2t}}$$

is the probability density of $W(t)$ and find the expectation and variance of $W(t)$.

Hint The density of $W(t)$ can be obtained from condition 3) of Definition 6.9 written for a single time t and a single Borel set. You will need the formula

$$\int_{-\infty}^{+\infty} e^{-\frac{x^2}{2}} dx = \sqrt{2\pi}$$

to compute the integrals in the expressions for the expectation and variance.

Remark 6.1

The results of Exercise 6.19 mean that $W(t)$ has the normal distribution with mean 0 and variance t.

Exercise 6.20

Show that

$$E(W(s)W(t)) = \min\{s, t\}.$$

Hint The joint density of $W(s)$ and $W(t)$ will be needed. It can be found from condition 3) of Definition 6.9 written for two times s and t and two Borel sets.

Exercise 6.21

Show that

$$E\left(|W(t) - W(s)|^2\right) = |t - s|.$$

Hint Expand the square and use the formula in Exercise 6.20.

Exercise 6.22

Compute the characteristic function $E(\exp(i\lambda W(t)))$ for any $\lambda \in \mathbb{R}$.

Hint Use the density of $W(t)$ found in Exercise 6.19.

Exercise 6.23

Find $E\left(W(t)^4\right)$.

Hint This can be done, for example, by expressing the expectation in terms of the density of $W(t)$ and computing the resulting integral, or by computing the fourth derivative of the characteristic function of $W(t)$ at 0. The second method is more efficient.

Definition 6.10

We call $W(t) = \left(W^1(t), \ldots, W^n(t)\right)$ an *n-dimensional Wiener process* if $W^1(t), \ldots, W^n(t)$ are independent $\mathbb{R}$-valued Wiener processes.

Exercise 6.24

For a two-dimensional Wiener process $W(t) = \left(W^1(t), W^2(t)\right)$ find the probability that $|W(t)| < R$, where $R > 0$ and $|x|$ is the Euclidean norm of $x = \left(x^1, x^2\right)$ in $\mathbb{R}^2$, i.e. $|x|^2 = \left(x^1\right)^2 + \left(x^2\right)^2$.

Hint Express the probability in terms of the joint density of $W^1(t)$ and $W^2(t)$. Independence means that the joint density of $W^1(t)$ and $W^2(t)$ is the product of their respective densities, which are known from Exercise 6.19. It is convenient to use polar coordinates to compute the resulting integral over a disc.

6.3.2 Increments of Brownian Motion

Proposition 6.2

For any $0 \leq s < t$ the increment $W(t) - W(s)$ has the normal distribution with mean 0 and variance $t - s$.

Proof

By condition 3) of Definition 6.9 the joint density of $W(s), W(t)$ is

$$f_{W(s),W(t)}(x,y) = p(s,0,x)\, p(t-s,x,y).$$

Hence, for any Borel set A

$$\begin{aligned}
P\{W(t) - W(s) \in A\} &= \int_{\{(x,y):x-y\in A\}} p(s,0,x)\, p(t-s,x,y)\, dx\, dy \\
&= \int_{-\infty}^{+\infty} p(s,0,x) \left(\int_{\{y:x-y\in A\}} p(t-s,x,y)\, dy \right) dx \\
&= \int_{-\infty}^{+\infty} p(s,0,x) \left(\int_A p(t-s,x,x-u)\, du \right) dx \\
&= \int_{-\infty}^{+\infty} p(s,0,x) \left(\int_A p(t-s,0,u)\, du \right) dx \\
&= \int_A p(t-s,0,u)\, du \int_{-\infty}^{+\infty} p(s,0,x)\, dx \\
&= \int_A p(t-s,0,u)\, du.
\end{aligned}$$

But $f(u) = p(t-s,0,u)$ is the density of the normal distribution with mean 0 and variance $t-s$, which proves the claim. □

Corollary 6.1

Proposition 6.2 implies that $W(t)$ has stationary increments.

Proposition 6.3

For any $0 = t_0 \le t_1 \le \cdots \le t_n$ the increments

$$W(t_1) - W(t_0), \ldots, W(t_n) - W(t_{n-1})$$

are independent.

Proof

From Proposition 6.2 we know that the increments of $W(t)$ have the normal distribution. Because normally distributed random variables are independent if and only if they are uncorrelated, it suffices to verify that

$$E[(W(u) - W(t))(W(s) - W(r))] = 0$$

for any $0 \leq r \leq s \leq t \leq u$. But this follows immediately from Exercise 6.20:

$$\begin{aligned} E\left[(W(u)-W(t))(W(s)-W(r))\right] &= E(W(u)W(s)) - E(W(u)W(r)) \\ &\quad - E(W(t)W(s)) + E(W(t)W(r)) \\ &= s - r - s + r \\ &= 0, \end{aligned}$$

as required. □

Corollary 6.2

For any $0 \leq s < t$ the increment $W(t) - W(s)$ is independent of the σ-field

$$\mathcal{F}_s = \sigma\left\{W(r) : 0 \leq r \leq s\right\}.$$

Proof

By Proposition 6.3 the random variables $W(t)-W(s)$ and $W(r)-W(0) = W(r)$ are independent if $0 \leq r \leq s \leq t$. Because the σ-field $\mathcal{F}_s$ is generated by such $W(r)$, it follows that $W(t) - W(s)$ is independent of $\mathcal{F}_s$. □

Exercise 6.25

Show that $W(t)$ is a martingale with respect to the filtration $\mathcal{F}_t$.

Hint Take advantage of the fact that $W(t) - W(s)$ is independent of $\mathcal{F}_s$ if $s < t$.

Exercise 6.26

Show that $|W(t)|^2 - t$ is a martingale with respect to the filtration $\mathcal{F}_t$.

Hint Once again, use the fact that $W(t) - W(s)$ is independent of $\mathcal{F}_s$ if $s < t$.

Let us state without proof the following useful characterization of the Wiener process in terms of its increments.

Theorem 6.3

A stochastic process $W(t), t \geq 0$, is a Wiener process if and only if the following conditions hold:

1) $W(0) = 0$ a.s.;
2) the sample paths $t \mapsto W(t)$ are continuous a.s.;

3) $W(t)$ has stationary independent increments;

4) the increment $W(t) - W(s)$ has the normal distribution with mean 0 and variance $t - s$ for any $0 \leq s < t$.

Exercise 6.27

Show that for any $T > 0$

$$V(t) = W(t + T) - W(T)$$

is a Wiener process if $W(t)$ is.

Hint Are the increments of $V(t)$ independent? What is their distribution? Does $V(t)$ have continuous paths? Is it true that $V(0) = 0$?

The Wiener process can also be characterized by its martingale properties. The following theorem is also given without proof.

Theorem 6.4 (Lévy's martingale characterization)

Let $W(t), t \geq 0$, be a stochastic process and let $\mathcal{F}_t = \sigma(W_s, s \leq t)$ be the filtration generated by it. Then $W(t)$ is a Wiener process if and only if the following conditions hold:

1) $W(0) = 0$ a.s.;

2) the sample paths $t \mapsto W(t)$ are continuous a.s.;

3) $W(t)$ is a martingale with respect to the filtration $\mathcal{F}_t$;

4) $|W(t)|^2 - t$ is a martingale with respect to $\mathcal{F}_t$.

Exercise 6.28

Let $c > 0$. Show that $V(t) = \frac{1}{c} W(c^2 t)$ is a Wiener process if $W(t)$ is.

Hint Is $V(t)$ a martingale? With respect to which filtration? Is $|V(t)|^2 - t$ a martingale? Are the paths of $V(t)$ continuous? Is it true that $V(0) = 0$?

6.3.3 Sample Paths

Let

$$0 = t_0^n < t_1^n < \cdots < t_n^n = T,$$

where

$$t_i^n = \frac{iT}{n},$$

be a partition of the interval $[0,T]$ into n equal parts. We denote by

$$\Delta_i^n W = W(t_{i+1}^n) - W(t_i^n)$$

the corresponding increments of the Wiener process $W(t)$.

Exercise 6.29

Show that

$$\lim_{n\to\infty} \sum_{i=0}^{n-1} (\Delta_i^n W)^2 = T \quad \text{in } L^2.$$

Hint You need to show that

$$\lim_{n\to\infty} E\left(\left[\sum_{i=0}^{n-1} (\Delta_i^n W)^2 - T\right]^2\right) = 0.$$

Use the independence of increments to simplify the expectation. What are the expectations of $\Delta_i^n W$, $(\Delta_i^n W)^2$ and $(\Delta_i^n W)^4$?

The next theorem on the variation of the paths of $W(t)$ is a consequence of the result in Exercise 6.29. First, let us recall that the variation of a function is defined as follows.

Definition 6.11

The *variation* of a function $f : [0,T] \to \mathbb{R}$ is defined to be

$$\limsup_{\Delta t\to 0} \sum_{i=0}^{n-1} |f(t_{i+1}) - f(t_i)|,$$

where $t = (t_0, t_1, \dots, t_n)$ is a partition of $[0,T]$, i.e. $0 = t_0 < t_1 < \cdots < t_n = T$, and where

$$\Delta t = \max_{i=0,\dots,n-1} |t_{i+1} - t_i|.$$

Theorem 6.5

The variation of the paths of $W(t)$ is infinite a.s.

Proof

Consider the sequence of partitions $t^n = (t_0^n, t_1^n, \ldots, t_n^n)$ of $[0,T]$ into n equal parts. Then

$$\sum_{i=0}^{n-1} |\Delta_i^n W|^2 \leq \left(\max_{i=0,\ldots,n-1} |\Delta_i^n W| \right) \sum_{i=0}^{n-1} |\Delta_i^n W| .$$

Since the paths of $W(t)$ are a.s. continuous on $[0,T]$,

$$\lim_{n\to\infty} \left(\max_{i=0,\ldots,n-1} |\Delta_i^n W| \right) = 0 \quad \text{a.s.}$$

By Exercise 6.29 there is a subsequence $t^{n_k} = (t_0^{n_k}, t_1^{n_k}, \ldots, t_{n_k}^{n_k})$ of partitions such that

$$\lim_{k\to\infty} \sum_{i=0}^{n_k-1} |\Delta_i^{n_k} W|^2 = T \quad \text{a.s.}$$

This is because every sequence of random variables convergent in L^2 has a subsequence convergent a.s. It follows that

$$\lim_{k\to\infty} \sum_{i=0}^{n_k-1} |\Delta_i^{n_k} W| = \infty \quad \text{a.s.,}$$

while

$$\lim_{k\to\infty} \Delta t^{n_k} = \lim_{k\to\infty} \frac{T}{n_k} = 0,$$

which proves the theorem. □

Theorem 6.5 has important consequences for the theory of stochastic integrals presented in the next chapter. This is because an integral of the form

$$\int_0^T f(t)\, dW(t)$$

cannot be defined pathwise (that is, separately for each $\omega \in \Omega$) as the Riemann–Stieltjes integral if the paths have infinite variation. It turns out that an intrinsically stochastic approach will be needed to tackle such integrals, see Chapter 7.

Exercise 6.30

Show that $W(t)$ is a.s. non-differentiable at $t = 0$.

Hint By Exercise 6.28 $V_c(t) = \frac{1}{c}W(c^2t)$ is a Wiener process for any $c > 0$. Deduce that the probability

$$P\left\{\frac{|W(t)|}{t} > cM \text{ for some } t \in [0, \tfrac{1}{c^2}]\right\}$$

is the same for each $c > 0$. What is the probability that the limit of $\frac{|W(t)|}{t}$ exists as $t \searrow 0$, then?

Exercise 6.31

Show that for any $t \geq 0$ the Wiener process $W(t)$ is a.s. non-differentiable at t.

Hint $V_t(s) = W(s+t) - W(t)$ is a Wiener process for any $t \geq 0$.

A weak point in the assertion in Exercise 6.31 is that for each t the event of measure 1 in which $W(t)$ is non-differentiable at t may turn out to be different for each $t \geq 0$. The theorem below, which is presented without proof, shows that in fact the same event of measure 1 can be chosen for each $t \geq 0$. This is not a trivial conclusion because the set of $t \geq 0$ is uncountable.

Theorem 6.6

With probability 1 the Wiener process $W(t)$ is non-differentiable at any $t \geq 0$.

6.3.4 Doob's Maximal L^2 Inequality for Brownian Motion

The inequality proved in this section is necessary to study the properties of stochastic integrals in the next chapter. It can be viewed as an extension of Doob's maximal L^2 inequality in Theorem 4.1 to the case of continuous time. In fact, in the result below the Wiener process can be replaced by any square integrable martingale $\xi(t)$, $t \geq 0$ with a.s. continuous paths.

Theorem 6.7 (Doob's maximal L^2 inequality)

For any $t > 0$

$$E\left(\max_{s \leq t} |W(s)|^2\right) \leq 4E|W(t)|^2. \tag{6.6}$$

Proof

For $t > 0$ and $n \in \mathbb{N}$ we define

$$M_k^n = \left|W\left(\frac{kt}{2^n}\right)\right|, \quad 0 \leq k \leq 2^n. \tag{6.7}$$

Then, by Jensen's inequality, $M_k^n, k = 0, \cdots, 2^n$, is a non-negative square integrable submartingale with respect to the filtration $\mathcal{F}_k^n = \mathcal{F}_{\frac{kt}{2^n}}$, so by Theorem 4.1

$$E\left(\max_{k\leq 2^n}|M_k^n|^2\right) \leq 4E|M_{2^n}^n|^2 = 4E|W(t)|^2.$$

Since $W(t)$ has a.s. continuous paths,

$$\lim_{n\to\infty}\max_{k\leq 2^n}|M_k^n|^2 = \max_{s\leq t}|W(s)|^2 \quad \text{a.s.}$$

Moreover, since $M_k^n = M_{2k}^{n+1}$, the sequence $\sup_{k\leq 2^n}|M_k^n|^2, n\in\mathbb{N}$, is increasing. Hence by the Lebesgue monotone convergence theorem $\max_{s\leq t}|W(s)|^2$ is an integrable function and

$$E\left(\max_{s\leq t}|W(s)|^2\right) = \lim_{n\to\infty}E\left(\max_{k\leq 2^n}|M_k^n|^2\right) \leq 4E|W(t)|^2,$$

completing the proof. □

6.3.5 Various Exercises on Brownian Motion

Exercise 6.32

Verify that the transition density $p(t, x, y)$ satisfies the diffusion equation

$$\frac{\partial p}{\partial t} = \frac{1}{2}\frac{\partial^2 p}{\partial y^2}.$$

Hint Simply differentiate the expression (6.5) for the transition density.

Exercise 6.33

Show that $Z(t) = -W(t)$ is a Wiener process if $W(t)$ is.

Hint Are the increments of $Z(t)$ independent? How are they distributed? Are the paths of $Z(t)$ continuous? Is it true that $Z(0) = 0$?

Exercise 6.34

Show that for any $0 \leq s < t$

$$P\{W(t)\in A|W(s)\} = \int_A p(t-s, W(s), y)\,dy.$$

Hint Write the conditional probability as the conditional expectation of $1_A(W(t))$ given $W(s)$. Compute the conditional expectation by transforming the integral of $1_A(W(t))$ over any event in the σ-field generated by $W(s)$. This can be done using the joint density of $W(s)$ and $W(t)$. Refer to the chapter on conditional expectation if necessary.

Exercise 6.35

Show that $e^{W(t)}e^{-\frac{t}{2}}$ is a martingale. (It is called the *exponential martingale.*)

Hint What is the expectation of $e^{W(t)-W(s)}$ for $s < t$? By independence it is equal to the conditional expectation of $e^{W(t)-W(s)}$ given $\mathcal{F}_s$. This will give the martingale condition.

Exercise 6.36

Compute $E(W(s)|W(t))$ for $0 \leq s < t$.

Hint You want to find a Borel function F such that $E(W(s)|W(t)) = F(W(t))$, i.e.

$$\int_{\{W(t)\in A\}} W(s)\, dP = \int_{\{W(t)\in A\}} F(W(t))\, dP.$$

Either side of this equality can be transformed using the joint density of $W(t)$ and $W(s)$.

6.4 Solutions

Solution 6.1

Suppose that η is a random variable with exponential distribution of rate λ. The distribution function of η is

$$F(t) = P\{\eta \leq t\} = 1 - P\{\eta > t\} = \begin{cases} 0 & \text{if } t < 0, \\ 1 - e^{-\lambda t} & \text{if } t \geq 0. \end{cases}$$

Therefore η has density

$$f(t) = \frac{d}{dt}F(t) = \begin{cases} 0 & \text{if } t < 0, \\ \lambda e^{-\lambda t} & \text{if } t > 0. \end{cases}$$

The distribution function $F(t)$ and density $f(t)$ are shown in Figure 6.6.

Solution 6.2

Using the density $f(t) = \lambda e^{-\lambda t}$ found in Exercise 6.1 and integrating by parts,

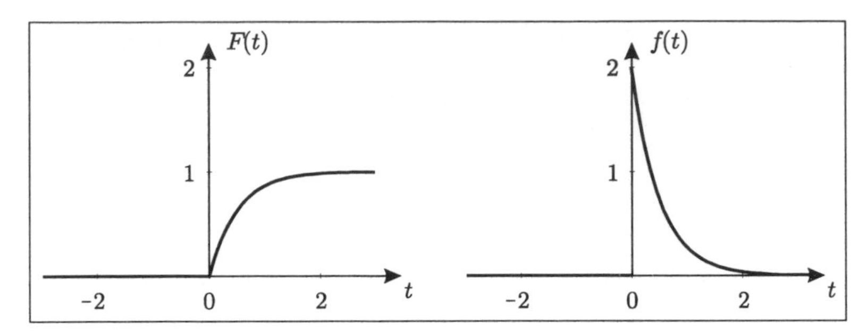

Figure 6.6. The distribution function $F(t)$ and density $f(t)$ of a random variable with exponential distribution of rate $\lambda = 2$

we obtain

$$\begin{aligned} E(\eta) &= \int_{-\infty}^{\infty} t f(t)\, dt = \int_0^{\infty} t\lambda e^{-\lambda t}\, dt = -\int_0^{\infty} t \frac{d}{dt} e^{-\lambda t}\, dt \\ &= -te^{-\lambda t}\Big|_0^{\infty} + \int_0^{\infty} e^{-\lambda t}\, dt = 0 - \frac{1}{\lambda} e^{-\lambda t}\Big|_0^{\infty} = \frac{1}{\lambda}. \end{aligned}$$

In a similar way we compute

$$\begin{aligned} E(\eta^2) &= \int_{-\infty}^{\infty} t^2 f(t)\, dt = \int_0^{\infty} t^2 \lambda e^{-\lambda t}\, dt = -\int_0^{\infty} t^2 \frac{d}{dt} e^{-\lambda t}\, dt \\ &= -t^2 e^{-\lambda t}\Big|_0^{\infty} + 2\int_0^{\infty} te^{-\lambda t}\, dt = 0 + \frac{2}{\lambda}\int_0^{\infty} t f(t)\, dt = \frac{2}{\lambda^2}. \end{aligned}$$

It follows that the variance is equal to

$$\operatorname{var}(\eta) = E\left(\eta^2\right) - \left(E(\eta)\right)^2 = \frac{2}{\lambda^2} - \frac{1}{\lambda^2} = \frac{1}{\lambda^2}.$$

Solution 6.3

By the definition of a random variable with exponential distribution

$$P\{\eta > s + t\} = e^{-\lambda(s+t)} = e^{-\lambda s} e^{-\lambda t} = P\{\eta > s\} P\{\eta > t\}.$$

Solution 6.4

By the definition of conditional probability

$$\begin{aligned} P\{\eta > t + s | \eta > s\} &= \frac{P\{\eta > t + s, \eta > s\}}{P\{\eta > s\}} \\ &= \frac{P\{\eta > t + s\}}{P\{\eta > s\}}, \end{aligned}$$

since $\{\eta > t + s, \eta > s\} = \{\eta > t + s\}$ (because $\eta > s + t$ implies that $\eta > s$). Now it is clear that (6.1) is equivalent to (6.2).

Solution 6.5

The exponential distribution is the only probability distribution satisfying the lack of memory property because the only non-negative non-increasing solutions of the functional equation

$$g(t+s) = g(t)g(s)$$

are of the form $g(t) = a^t$ for some $0 \leq a \leq 1$.

To verify this observe that $[g(m/n)]^n = g(m) = [g(1)]^m$ for any integers m and $n \neq 0$. Let $a := g(1)$. It follows that

$$g(q) = a^q \quad \text{for any } q \in \mathbb{Q}.$$

Since g is non-increasing, $0 \leq a \leq 1$ and

$$a^t = \inf_{t > q \in \mathbb{Q}} g(q) \geq g(t) \geq \sup_{t < q \in \mathbb{Q}} g(q) = a^t,$$

so indeed

$$g(t) = a^t \quad \text{for any } t \in \mathbb{R}.$$

As a result, $P\{\eta > t\} = a^t$ for some $0 \leq a \leq 1$. But the distribution function of a random variable cannot be constant, so $0 \neq a \neq 1$. Hence $a = e^{-\lambda}$ for some $\lambda > 0$, completing the argument.

Solution 6.6

Since $N(t)$ has the Poisson distribution with parameter λt we have $E(N(t)) = \lambda t$. Indeed

$$\begin{aligned} E(N(t)) &= \sum_{n=0}^{\infty} nP\{N(t) = n\} = \sum_{n=0}^{\infty} ne^{-\lambda t} \frac{(\lambda t)^n}{n!} \\ &= \lambda t e^{-\lambda t} \sum_{n=1}^{\infty} \frac{(\lambda t)^{n-1}}{(n-1)!} = \lambda t e^{-\lambda t} e^{\lambda t} = \lambda t. \end{aligned}$$

Solution 6.7

Using the fact that $\eta_1, \eta_2, \ldots$ are independent and exponentially distributed, we obtain

$$\begin{aligned} P\{N(s) = 1, N(t) = 2\} &= P\{\xi_1 \leq s < \xi_2 \leq t < \xi_3\} \\ &= P\{\eta_1 \leq s < \eta_1 + \eta_2 \leq t < \eta_1 + \eta_2 + \eta_3\} \\ &= \int_0^s P\{s < u + \eta_2 \leq t < u + \eta_2 + \eta_3\} \lambda e^{-\lambda u} du \\ &= \int_0^s \left(\int_{s-u}^{t-u} P\{t < u + v + \eta_3\} \lambda e^{-\lambda v} dv \right) \lambda e^{-\lambda u} du \end{aligned}$$

$$
\begin{aligned}
&= \int_0^s \left(\int_{s-u}^{t-u} e^{-\lambda(t-u-v)} \lambda e^{-\lambda v} dv \right) \lambda e^{-\lambda u} du \\
&= \lambda^2 e^{-\lambda t} s\,(t-s)\,.
\end{aligned}
$$

Solution 6.8

Since

$$
\begin{aligned}
\xi_n^t &= \eta_1^t + \cdots + \eta_n^t \\
&= \xi_{N(t)+1} - t + \eta_{N(t)+2} + \cdots + \eta_{N(t)+n} \\
&= \xi_{N(t)+n} - t,
\end{aligned}
$$

it follows that

$$
\begin{aligned}
N^t(s) &= \max\{n : \xi_n^t \le s\} \\
&= \max\{n : \xi_{N(t)+n} \le t+s\} \\
&= \max\{n : \xi_n \le t+s\} - N(t) \\
&= N(t+s) - N(t).
\end{aligned}
$$

Solution 6.9

It is easily verified that

$$
\begin{aligned}
\{N(t) = n\} &= \{\eta_{n+1} > t - \xi_n, t \ge \xi_n\} \\
\{\eta_1^t > s, N(t) = n\} &= \{\eta_{n+1} > s + t - \xi_n, t \ge \xi_n\}
\end{aligned}
$$

Since ξ_n, η_{n+1} are independent and η_{n+1} satisfies the lack of memory property from Exercise 6.3,

$$
\begin{aligned}
P\{\eta_1^t > s, N(t) = n\} &= P\{\eta_{n+1} > s + t - \xi_n, t \ge \xi_n\} \\
&= \int_{-\infty}^t P\{\eta_{n+1} > s + t - u\} P_{\xi_n}(du) \\
&= P\{\eta_{n+1} > s\} \int_{-\infty}^t P\{\eta_{n+1} > t - u\} P_{\xi_n}(du) \\
&= P\{\eta_{n+1} > s\} P\{\eta_{n+1} > t - \xi_n, t \ge \xi_n\} \\
&= P\{\eta_{n+1} > s\} P\{N(t) = n\} \\
&= P\{\eta_1 > s\} P\{N(t) = n\}\,.
\end{aligned}
$$

The last equality holds because η_{n+1} has the same distribution as η_1. Now divide both sides by $P\{N(t) = n\}$ to get

$$
P\{\eta_1^t > s | N(t) = n\} = P\{\eta_1 > s\}
$$

for any $n = 0, 1, 2, \ldots$. As a result,

$$P\{\eta_1^t > s | N(t)\} = P\{\eta_1 > s\}$$

because $N(t)$ is a discrete random variable with values $0, 1, 2, \ldots$.

Solution 6.10

As in Solution 6.9,

$$\begin{aligned} \{N(t) = n\} &= \{\eta_{n+1} > t - \xi_n, t \geq \xi_n\}, \\ \{\eta_1^t > s_1, N(t) = n\} &= \{\eta_{n+1} > s_1 + t - \xi_n, t \geq \xi_n\} \end{aligned}$$

and, more generally,

$$\begin{aligned} &\{\eta_1^t > s_1, \eta_2^t > s_2, \ldots, \eta_k^t > s_k, N(t) = n\} \\ &= \{\eta_{n+1} > s_1 + t - \xi_n, t \geq \xi_n\} \cap \{\eta_{n+2} > s_2\} \cap \cdots \cap \{\eta_{n+k} > s_k\}. \end{aligned}$$

Since $\xi_n, \eta_{n+1}, \ldots, \eta_{n+k}$ are independent and $\eta_{n+2}, \ldots, \eta_{n+k}$ have the same distribution as $\eta_2, \ldots, \eta_k$, using Exercise 6.9 we find that

$$\begin{aligned} &P\{\eta_1^t > s_1, \eta_2^t > s_2, \ldots, \eta_k^t > s_k, N(t) = n\} \\ &= P\{\eta_{n+1} > s_1 + t - \xi_n, \xi_n \leq t\} P\{\eta_{n+2} > s_2\} \cdots P\{\eta_{n+k} > s_k\} \\ &= P\{\eta_1^t > s_1, N(t) = n\} P\{\eta_2 > s_2\} \cdots P\{\eta_k > s_k\} \\ &= P\{\eta_1^t > s_1 | N(t) = n\} P\{\eta_2 > s_2\} \cdots P\{\eta_k > s_k\} P\{N(t) = n\} \\ &= P\{\eta_1 > s\} P\{\eta_2 > s_2\} \cdots P\{\eta_k > s_k\} P\{N(t) = n\}. \end{aligned}$$

As in Solution 6.9, this implies the desired equality.

Solution 6.11

Take the expectation on both sides of the equality in Exercise 6.10 to find that

$$P\{\eta_1^t > s_1, \ldots, \eta_k^t > s_k\} = P\{\eta_1 > s_1\} \cdots P\{\eta_k > s_k\}.$$

If all the numbers s_n except perhaps one are zero, it follows that

$$P\{\eta_n^t > s_n\} = P\{\eta_n > s_n\}, \quad n = 1, \ldots, k,$$

so the random variables η_n^t have the same distribution as η_n. Inserting this back into the first equality, we obtain

$$P\{\eta_1^t > s_1, \ldots, \eta_k^t > s_k\} = P\{\eta_1^t > s_1\} \cdots P\{\eta_k^t > s_k\},$$

so the η_n^t are independent.

To prove that the η_n^t are independent of $N(t)$ integrate the formula in Exercise 6.10 over $\{N(t) = n\}$ and multiply by $P\{N(t) = n\}$ to get

$$P\{\eta_1^t > s_1, \ldots, \eta_k^t > s_k, N(t) = n\} = P\{\eta_1 > s_1\} \cdots P\{\eta_k > s_k\} P\{N(t) = n\}.$$

But $P\{\eta_n^t > s_n\} = P\{\eta_n > s_n\}$, hence $N(t)$ and the η_n^t are independent.

Solution 6.12

We need to verify conditions 1), 2), 3) of Definition 6.3. Clearly, $N(t) - \lambda t$ is $\mathcal{F}_t$-measurable. By Exercise 6.6

$$E(|N(t)|) = E(N(t)) = \lambda t < \infty,$$

which means that $N(t)$ is integrable, and so is $N(t) - \lambda t$.

Theorem 6.2 implies that $N(t) - N(s)$ is independent of $\mathcal{F}_s$ for any $0 \leq s \leq t$, so

$$E(N(t) - N(s)|\mathcal{F}_s) = E(N(t) - N(s)) = E(N(t)) - E(N(s)) = \lambda t - \lambda s.$$

It follows that

$$E(N(t) - \lambda t|\mathcal{F}_s) = E(N(s) - \lambda s|\mathcal{F}_s) = N(s) - \lambda s,$$

completing the proof.

Solution 6.13

Since $\eta_n = \xi_n - \xi_{n-1}$ and $P\{\eta_n > 0\} = e^0 = 1$,

$$P\{\xi_0 < \xi_1 < \xi_2 < \cdots\} = P\left(\bigcap_{n=1}^{\infty}\{\eta_n > 0\}\right) = 1.$$

Here we have used the property that if $P(A_n) = 1$ for all $n = 1, 2, \ldots$, then $P\left(\bigcap_{n=1}^{\infty} A_n\right) = 1$.

Solution 6.14

Since

$$\lim_{n\to\infty} \xi_n = \sum_{n=1}^{\infty} \eta_n,$$

it follows that

$$\begin{aligned}\left\{\lim_{n\to\infty} \xi_n < \infty\right\} &\subset \{\eta_1, \eta_2, \ldots \text{ is a bounded sequence}\} \\ &= \bigcup_{m=1}^{\infty}\bigcap_{n=1}^{\infty}\{\eta_n \leq m\}.\end{aligned}$$

Let us compute the probability of this event. Because $\bigcap_{n=1}^{N}\{\eta_n \leq m\}$, $N = 1, 2, \ldots$ is a contracting sequence of events,

$$P\left(\bigcap_{n=1}^{\infty}\{\eta_n \leq m\}\right) = \lim_{N\to\infty} P\left(\bigcap_{n=1}^{N}\{\eta_n \leq m\}\right)$$

$$
\begin{aligned}
&= \lim_{N\to\infty} \prod_{n=1}^{N} P\{\eta_n \le m\} \\
&= \lim_{N\to\infty} \left(1 - e^{-\lambda m}\right)^N \\
&= 0.
\end{aligned}
$$

It follows that

$$
\begin{aligned}
P\left(\lim_{n\to\infty} \xi_n < \infty\right) &\le P\left(\bigcup_{m=1}^{\infty}\bigcap_{n=1}^{\infty}\{\eta_n \le m\}\right) \\
&\le \sum_{m=1}^{\infty} P\left(\bigcap_{n=1}^{\infty}\{\eta_n \le m\}\right) \\
&= 0,
\end{aligned}
$$

completing the proof.

While it is instructive to work through the above estimates, there exists a much more elegant argument. By the strong law of large numbers

$$
\lim_{n\to\infty} \frac{\xi_n}{n} = \frac{1}{\lambda} \quad \text{a.s.}
$$

Here $\frac{1}{\lambda}$ is the expectation of each of the independent identically distributed random variables η_n (see Exercise 6.2). It follows that

$$
\lim_{n\to\infty} \xi_n = \infty \quad \text{a.s.},
$$

as required.

Solution 6.15

In the proof of Proposition 6.1 it was shown that

$$
P\{\xi_n > t\} = e^{-\lambda t} \sum_{k=0}^{n-1} \frac{(\lambda t)^k}{k!}
$$

for $t \ge 0$, see (6.4). Therefore the distribution function

$$
F_n(t) = P\{\xi_n \le t\} = 1 - P\{\xi_n > t\} = e^{-\lambda t} \sum_{k=n}^{\infty} \frac{(\lambda t)^k}{k!}
$$

of ξ_n is differentiable, the density f_n of ξ_n being

$$
\begin{aligned}
f_n(t) &= \frac{d}{dt} F_n(t) \\
&= -\lambda e^{-\lambda t} \sum_{k=n}^{\infty} \frac{(\lambda t)^k}{k!} + \lambda e^{-\lambda t} \sum_{k=n}^{\infty} \frac{(\lambda t)^{k-1}}{(k-1)!} \\
&= \lambda e^{-\lambda t} \frac{(\lambda t)^{n-1}}{(n-1)!}
\end{aligned}
$$

for $t > 0$, and clearly $f_n(t) = 0$ for $t \leq 0$.

Solution 6.16

Because $N(t)$ has non-decreasing trajectories

$$\left\{\lim_{t\to\infty} N(t) = \infty\right\} = \bigcap_{n=1}^{\infty}\bigcup_{k=1}^{\infty}\{N(k) \geq n\}.$$

Also, $\{N(k) \geq n\}$, $k = 1, 2, \ldots$ is an expanding sequence of events and

$$\begin{aligned} P\{N(k) \geq n\} &= e^{-\lambda k}\sum_{i=n}^{\infty}\frac{(\lambda k)^i}{i!} \\ &= 1 - e^{-\lambda k}\sum_{i=0}^{n-1}\frac{(\lambda k)^i}{i!} \to 1 \quad \text{as } k \to \infty. \end{aligned}$$

It follows that

$$P\left\{\bigcup_{k=1}^{\infty}\{N(k) \geq n\}\right\} = 1,$$

so

$$P\left\{\lim_{t\to\infty} N(t) = \infty\right\} = P\left\{\bigcap_{n=1}^{\infty}\bigcup_{k=1}^{\infty}\{N(k) \geq n\}\right\} = 1.$$

Solution 6.17

Since

$$\begin{aligned} \sinh(x) &= \frac{e^x - e^{-x}}{2} = \sum_{n=0}^{\infty}\frac{x^{2n+1}}{(2n+1)!}, \\ \cosh(x) &= \frac{e^x + e^{-x}}{2} = \sum_{n=0}^{\infty}\frac{x^{2n}}{(2n)!}, \end{aligned}$$

we have

$$\begin{aligned} P\{N(t) \text{ is odd}\} &= \sum_{n=0}^{\infty} P\{N(t) = 2n+1\} \\ &= \sum_{n=0}^{\infty} e^{-\lambda t}\frac{(\lambda t)^{2n+1}}{(2n+1)!} \\ &= e^{-\lambda t}\sinh(\lambda t), \\ P\{N(t) \text{ is even}\} &= \sum_{n=0}^{\infty} P\{N(t) = 2n\} \\ &= \sum_{n=0}^{\infty} e^{-\lambda t}\frac{(\lambda t)^{2n}}{(2n)!} \\ &= e^{-\lambda t}\cosh(\lambda t). \end{aligned}$$

Solution 6.18

We can write

$$N(n) = N(1) + (N(2) - N(1)) + \cdots + (N(n) - N(n-1)),$$

where $N(1), N(2) - N(1), N(3) - N(2), \ldots$ is a sequence of independent identically distributed random variables with expectation

$$E(N(1)) = E(N(2) - N(1)) = E(N(3) - N(2)) = \cdots = \lambda.$$

By the strong law of large numbers

$$\lim_{n\to\infty} \frac{N(n)}{n} = \lambda \quad \text{a.s.} \tag{6.8}$$

Now, if $n \leq t \leq n+1$, then $N(n) \leq N(t) \leq N(n+1)$ and

$$\frac{N(n)}{n+1} \leq \frac{N(t)}{t} \leq \frac{N(n+1)}{n}.$$

By (6.8) both sides tend to λ as $n \to \infty$, implying that

$$\lim_{t\to\infty} \frac{N(t)}{t} = \lambda \quad \text{a.s.}$$

Solution 6.19

Condition 3) of Definition 6.9 implies that

$$f_{W(t)}(x) = p(t, 0, x)$$

is the density of $W(t)$. Therefore, integrating by parts, we can compute the expectation

$$\begin{aligned} E(W(t)) &= \int_{-\infty}^{+\infty} x\, p(t,0,x)\, dx \\ &= \frac{1}{\sqrt{2\pi t}} \int_{-\infty}^{+\infty} x\, e^{-\frac{x^2}{2t}}\, dx \\ &= -\frac{t}{\sqrt{2\pi t}} \int_{-\infty}^{+\infty} \frac{d}{dx} e^{-\frac{x^2}{2t}}\, dx \\ &= -\frac{t}{\sqrt{2\pi t}}\, e^{-\frac{x^2}{2t}} \Big|_{-\infty}^{+\infty} = 0 \end{aligned}$$

and variance

$$\begin{aligned} E\left((W(t))^2\right) &= \int_{-\infty}^{+\infty} x^2\, p(t,0,x)\, dx \\ &= \frac{1}{\sqrt{2\pi t}} \int_{-\infty}^{+\infty} x^2\, e^{-\frac{x^2}{2t}}\, dx \\ &= -\frac{t}{\sqrt{2\pi t}} \int_{-\infty}^{+\infty} x \frac{d}{dx} e^{-\frac{x^2}{2t}}\, dx \\ &= -\frac{t}{\sqrt{2\pi t}}\, x e^{-\frac{x^2}{2t}} \Big|_{-\infty}^{+\infty} + \frac{t}{\sqrt{2\pi t}} \int_{-\infty}^{+\infty} e^{-\frac{x^2}{2t}}\, dx \\ &= 0 + \frac{t}{\sqrt{2\pi}} \int_{-\infty}^{+\infty} e^{-\frac{u^2}{2}}\, du = t. \end{aligned}$$

We have used the substitution $u = \frac{x}{\sqrt{t}}$ and the formula stated in the hint.

Solution 6.20

Suppose that $s < t$. Condition 3) of Definition 6.9 implies that the joint density of $W(s)$ and $W(t)$ is

$$f_{W(s),W(t)}(x,y) = p(s,0,x)\, p(t-s,x,y).$$

It follows that

$$\begin{aligned} E(W(s)W(t)) &= \int_{-\infty}^{+\infty} \int_{-\infty}^{+\infty} xy\; p(s,0,x)\, p(t-s,x,y)\, dx\, dy \\ &= \int_{-\infty}^{+\infty} x\, p(s,0,x) \left(\int_{-\infty}^{+\infty} y\, p(t-s,x,y)\, dy \right) dx \\ &= \int_{-\infty}^{+\infty} x^2 p(s,0,x)\, dx = s. \end{aligned}$$

This is because by the results of Exercise 6.19

$$\begin{aligned} \int_{-\infty}^{+\infty} y\, p(t-s,x,y)\, dy &= \int_{-\infty}^{+\infty} (x+u)\, p(t-s,x,x+u)\, du \\ &= \int_{-\infty}^{+\infty} (x+u)\, p(t-s,0,u)\, du \\ &= x \int_{-\infty}^{+\infty} p(t-s,0,u)\, du + \int_{-\infty}^{+\infty} u\, p(t-s,0,u)\, du \\ &= x + 0 = x \end{aligned}$$

and

$$\int_{-\infty}^{+\infty} x^2 p(s,0,x)\, dx = s.$$

It follows that for arbitrary $s, t \geq 0$

$$E(W(s)W(t)) = \min\{s,t\}.$$

Solution 6.21

Suppose that $s \leq t$. Then by Exercise 6.20

$$\begin{aligned} E\left(|W(t)-W(s)|^2\right) &= E\left(W(t)^2\right) - 2E(W(s)W(t)) + E\left(W(s)^2\right) \\ &= t - 2s + s = t - s. \end{aligned}$$

In general, for arbitrary $s, t \geq 0$

$$E\left(|W(t)-W(s)|^2\right) = |t-s|.$$

Solution 6.22

Using the density $f_{W(t)}(x) = p(t,0,x)$ of $W(t)$, we compute

$$\begin{aligned} E(\exp(i\lambda W(t))) &= \int_{-\infty}^{+\infty} e^{i\lambda x} p(t,0,x)\, dx \\ &= \frac{1}{\sqrt{2\pi t}} \int_{-\infty}^{+\infty} e^{i\lambda x} e^{-\frac{x^2}{2t}}\, dx \\ &= \frac{1}{\sqrt{2\pi t}} e^{-\frac{\lambda^2 t}{2}} \int_{-\infty}^{+\infty} e^{-\frac{(x-i\lambda t)^2}{2t}}\, dx \\ &= e^{-\frac{\lambda^2 t}{2}}. \end{aligned}$$

Solution 6.23

Using the formula for the characteristic function of $W(t)$ found in Exercise 6.22, we compute

$$\begin{aligned} E\left(W(t)^4\right) &= \left.\frac{d^4}{d\lambda^4}\right|_{\lambda=0} E(\exp(i\lambda W(t))) \\ &= \left.\frac{d^4}{d\lambda^4}\right|_{\lambda=0} e^{-\frac{1}{2}\lambda^2 t} \\ &= 3t^2. \end{aligned}$$

Solution 6.24

Since $W^1(t), W^2(t)$ are independent, their joint density is the product of the densities of $W^1(t)$ and $W^2(t)$. Therefore

$$P\{|W(t)| < R\} = \int_{\{|x|<R\}} p(t,0,x)\, p(t,0,y)\, dx\, dy$$

$$\begin{aligned}
&= \frac{1}{2\pi t}\int_{\{|x|<R\}} e^{-\frac{x^2+y^2}{2t}}\,dx\,dy \\
&= \frac{1}{2\pi t}\int_0^R\int_0^{2\pi} re^{-\frac{r^2}{2t}}\,d\varphi\,dr \\
&= -\int_0^R \frac{d}{dr}e^{-\frac{r^2}{2t}}\,dr \\
&= 1-e^{-\frac{R^2}{2t}}.
\end{aligned}$$

We have used the polar coordinates R,φ to compute the integral.

Solution 6.25

For any $0 \leq s < t$

$$\begin{aligned}
E\left(W(t)|\mathcal{F}_s\right) &= E\left(W(t)-W(s)|\mathcal{F}_s\right)+E\left(W(s)|\mathcal{F}_s\right) \\
&= E\left(W(t)-W(s)\right)+W(s) \\
&= W(s),
\end{aligned}$$

since $W(t)-W(s)$ is independent of $\mathcal{F}_s$ by Corollary 6.2, $W(s)$ is $\mathcal{F}_s$-measurable and $E\left(W(t)\right)=E\left(W(s)\right)=0$.

Solution 6.26

For any $0 \leq s < t$

$$\begin{aligned}
E\left(W(t)^2|\mathcal{F}_s\right) &= E\left(|W(t)-W(s)|^2|\mathcal{F}_s\right)+E\left(2W(t)W(s)|\mathcal{F}_s\right) \\
&\quad -E\left(W(s)^2|\mathcal{F}_s\right) \\
&= E\left(|W(t)-W(s)|^2\right)+2W(s)E\left(W(t)|\mathcal{F}_s\right) \\
&\quad -W(s)^2 \\
&= t-s+2W(s)^2-W(s)^2 \\
&= t-s+W(s)^2,
\end{aligned}$$

since $W(t)-W(s)$ is independent of $\mathcal{F}_s$ and has the normal distribution with mean 0 and variance $t-s$, $W(s)$ is $\mathcal{F}_s$-measurable, and $W(t)$ is a martingale. It follows that

$$E\left(W(t)^2-t|\mathcal{F}_s\right)=W(s)^2-s,$$

as required.

Solution 6.27

For any $0 \leq t_0 < t_1 < \cdots < t_n$ the increments

$$V(t_n)-V(t_{n-1}),\cdots,V(t_1)-V(t_0)$$

of $V(t)$ are independent, since the increments

$$W(t_n+T)-W(t_{n-1}+T),\cdots,W(t_1+T)-W(t_0+T)$$

of $W(t)$ are independent. For any $0 \leq s < t$ the increment $V(t)-V(s)$ has the normal distribution with mean zero and variance $t-s$, since $W(t+T)-W(s+T)$ does. Moreover, the paths $t \mapsto V(t) = W(t+T)$ are continuous and

$$V(0) = W(T) - W(T) = 0.$$

By Theorem 6.3 $V(t)$ is a Wiener process.

Solution 6.28

It is clear that $V(0) = \frac{1}{c}W(0) = 0$ a.s. and the paths $t \mapsto V(t) = \frac{1}{c}W(c^2 t)$ are a.s. continuous. We shall verify that $V(t)$ and $|V(t)|^2 - t$ are martingales with respect to the filtration

$$\begin{aligned}
\mathcal{G}_t &= \sigma\{V(s) : 0 \leq s \leq t\} \\
&= \sigma\left\{W(c^2 s) : 0 \leq s \leq t\right\} \\
&= \sigma\left\{W(s) : 0 \leq s \leq c^2 t\right\} \\
&= \mathcal{F}_{c^2 t}.
\end{aligned}$$

Indeed, if $s < t$, then $c^2 s < c^2 t$, so

$$\begin{aligned}
E\left(V(t)|\mathcal{G}_s\right) &= E\left(\frac{1}{c}W(c^2 t)|\mathcal{F}_{c^2 s}\right) \\
&= \frac{1}{c}E\left(W(c^2 t)|\mathcal{F}_{c^2 s}\right) \\
&= \frac{1}{c}W(c^2 s) = V(s)
\end{aligned}$$

and

$$\begin{aligned}
E\left(|V(t)|^2 - t|\mathcal{G}_s\right) &= E\left(\frac{1}{c^2}\left|W(c^2 t)\right|^2 - t|\mathcal{F}_{c^2 s}\right) \\
&= \frac{1}{c^2}E\left(\left|W(c^2 t)\right|^2 - c^2 t|\mathcal{F}_{c^2 s}\right) \\
&= \frac{1}{c^2}\left(\left|W(c^2 s)\right|^2 - c^2 s\right) \\
&= |V(s)|^2 - s,
\end{aligned}$$

since $W(t)$ and $|W(t)|^2 - t$ are martingales with respect to the filtration $\mathcal{F}_t$. It follows by Levy's martingale characterization that $V(t)$ is a Wiener process.

Solution 6.29

Since the increments $\Delta_i^n W$ are independent and

$$E(\Delta_i^n W) = 0, \quad E\left((\Delta_i^n W)^2\right) = \frac{T}{n}, \quad E\left((\Delta_i^n W)^4\right) = \frac{3T^2}{n^2},$$

it follows that

$$\begin{aligned}
E\left(\left[\sum_{i=0}^{n-1} (\Delta_i^n W)^2 - T\right]^2\right) &= E\left(\left[\sum_{i=0}^{n-1} \left((\Delta_i^n W)^2 - \frac{T}{n}\right)\right]^2\right) \\
&= \sum_{i=0}^{n-1} E\left[\left((\Delta_i^n W)^2 - \frac{T}{n}\right)^2\right] \\
&= \sum_{i=0}^{n-1} \left[E\left((\Delta_i^n W)^4\right) - \frac{2T}{n} E\left((\Delta_i^n W)^2\right) + \frac{T^2}{n^2}\right] \\
&= \sum_{i=0}^{n-1} \left[\frac{3T^2}{n^2} - \frac{2T^2}{n^2} + \frac{T^2}{n^2}\right] = \frac{2T^2}{n} \to 0
\end{aligned}$$

as $n \to \infty$.

Solution 6.30

We claim that, with probability 1, for any positive integer n there is a $t \in [0, \frac{1}{n^4}]$ such that $\frac{|W(t)|}{t} > n$. This condition implies that $W(t)$ is not differentiable at $t = 0$.

Let us put

$$A_n = \left\{ \tfrac{|W(t)|}{t} > n \text{ for some } t \in [0, \tfrac{1}{n^4}] \right\}$$

By Exercise 6.28

$$V_n(t) = \frac{1}{n^2} W(n^4 t)$$

is a Brownian motion for any n. Therefore

$$\begin{aligned}
P(A_n) &\geq P\left\{ \tfrac{|W(1/n^4)|}{1/n^4} > n \right\} \\
&= P\left\{ \tfrac{|V(1/n^4)|}{1/n^4} > n \right\} \\
&= P\left\{ |W(1)| > \tfrac{1}{n} \right\} \to 1 \quad \text{as } n \to \infty.
\end{aligned}$$

Since $A_1, A_2, \ldots$ is a contracting sequence of events,

$$P\left(\bigcap_{n=1}^{\infty} A_n \right) = \lim_{n \to \infty} P(A_n) = 1,$$

which proves the claim.

Solution 6.31

By Exercise 6.27 $V_t(s) = W(s+t) - W(t)$ is a Wiener process for any $t \geq 0$. Therefore, by Exercise 6.30 $V_t(s)$ is a.s. non-differentiable at $s = 0$. But this implies that $W(t)$ is a.s. non-differentiable at t.

Solution 6.32

Differentiating

$$p(t,x,y) = \frac{1}{\sqrt{2\pi t}} e^{-\frac{(y-x)^2}{2t}},$$

we obtain

$$\begin{aligned}
\frac{\partial}{\partial t} p(t,x,y) &= \frac{y^2 - 2yx + x^2 - t}{2t} p(t,x,y), \\
\frac{\partial}{\partial y} p(t,x,y) &= \frac{x-y}{t} p(t,x,y), \\
\frac{\partial^2}{\partial y^2} p(t,x,y) &= \frac{y^2 - 2yx + x^2 - t}{t} p(t,x,y),
\end{aligned}$$

so

$$\frac{\partial p}{\partial t} = \frac{1}{2} \frac{\partial^2 p}{\partial y^2},$$

as required.

Solution 6.33

Clearly, $Z(t) = -W(t)$ has a.s. continuous trajectories and $Z(0) = -W(0) = 0$ a.s. If $W(t)$ has stationary independent increments, then so does $Z(t) = -W(t)$. Finally,

$$Z(t) - Z(s) = -(W(t) - W(s))$$

has the same distribution as $W(t) - W(s)$, i.e. normal with mean 0 and variance $t - s$. By Theorem 6.3 $Z(t)$ is a Wiener process.

Solution 6.34

Let $0 \leq s < t$. Then

$$\begin{aligned}
\int_{\{W(s)\in B\}} 1_A(W(t))\, dP &= P\{W(s) \in B, W(t) \in A\} \\
&= \int_B \int_A p(s,0,x) p(t-s,x,y)\, dx\, dy \\
&= \int_B \left(\int_A p(t-s,x,y)\, dy \right) p(s,0,x)\, dx \\
&= \int_{\{W(s)\in B\}} \left(\int_A p(t-s,W(t),y)\, dy \right) dP
\end{aligned}$$

for any Borel set $B \subset \mathbb{R}$. It follows that

$$P\{W(t) \in A|W(s)\} = E(1_A(W(t))|W(s)) = \int_A p(t-s, W(t), y)\, dy.$$

Solution 6.35

We shall prove that $e^{W(t)}e^{-\frac{t}{2}}$ is a martingale with respect to the filtration $\mathcal{F}_t$. Clearly, it is adapted to the filtration $\mathcal{F}_t$, since $W(t)$ is. Let $0 \le s < t$. Because $W(t) - W(s)$ is independent of $\mathcal{F}_s$ and $W(s)$ is $\mathcal{F}_s$-measurable,

$$\begin{aligned} E\left(e^{W(t)}|\mathcal{F}_s\right) &= E\left(e^{W(t)-W(s)}e^{W(s)}|\mathcal{F}_s\right) \\ &= e^{W(s)}E\left(e^{W(t)-W(s)}|\mathcal{F}_s\right) \\ &= e^{W(s)}E\left(e^{W(t)-W(s)}\right). \end{aligned}$$

The increment $W(t) - W(s)$ has the normal distribution with mean 0 and variance $t - s$, so the expectation of $e^{W(t)-W(s)}$ is equal to

$$\begin{aligned} E\left(e^{W(t)-W(s)}\right) &= \int_{-\infty}^{+\infty} e^x p(t-s, 0, x)\, dx \\ &= e^{\frac{t-s}{2}} \int_{-\infty}^{+\infty} p(t-s, 0, x-t)\, dx \\ &= e^{\frac{t-s}{2}}. \end{aligned}$$

It follows that

$$E\left(e^{W(t)}e^{-\frac{t}{2}}|\mathcal{F}_s\right) = e^{W(s)}e^{-\frac{s}{2}}.$$

It also follows that $e^{W(t)}e^{-\frac{t}{2}}$ is integrable. Therefore $e^{W(t)}e^{-\frac{t}{2}}$ is a martingale.

Solution 6.36

Let $0 \le s < t$. We are looking for a Borel function F such that $E(W(s)|W(t)) = F(W(t))$, i.e.

$$\int_{\{W(t)\in A\}} W(s)\, dP = \int_{\{W(t)\in A\}} F(W(t))\, dP$$

for any Borel set A in $\mathbb{R}$. The integral on the right-hand side can be written as

$$\int_{\{W(t)\in A\}} F(W(t))\, dP = \int_A F(y)\, p(t, 0, y)\, dy$$

and the integral on the left-hand side as

$$\int_{\{W(t)\in A\}} W(s)\, dP = \int_A \left(\int_{-\infty}^{+\infty} x\, p(s, 0, x)\, p(t-s, x, y)\, dx\right) dy$$

using the expression for the joint density of $W(s)$ and $W(t)$ in Solution 6.20. Let us compute the inner integral:

$$\begin{aligned}\int_{-\infty}^{+\infty} xp(s,0,x)p(t-s,x,y)\,dx &= p(t,0,y)\int_{-\infty}^{\infty} xp\left(\frac{s\,(t-s)}{t},\frac{s}{t}y,x\right)dx \\ &= \frac{s}{t}y\,p(t,0,y).\end{aligned}$$

(To see that the first equality holds, just use formula (6.5) for $p(t,x,y)$.) Therefore

$$\int_{\{W(t)\in A\}} W(s)\,dP = \int_A \frac{s}{t}y\,p\,(t,0,y)\,dy.$$

It follows that $F(y) = \frac{s}{t}y$, i.e.

$$E\left(W(s)|W(t)\right) = \frac{s}{t}W(t).$$

7
Itô Stochastic Calculus

One of the first applications of the Wiener process was proposed by Bachelier, who around 1900 wrote a ground-breaking paper on the modelling of asset prices at the Paris Stock Exchange. Of course Bachelier could not have called it the Wiener process, but he used what in modern terminology amounts to $W(t)$ as a description of the market fluctuations affecting the price $X(t)$ of an asset. Namely, he assumed that infinitesimal price increments $dX(t)$ are proportional to the increments $dW(t)$ of the Wiener process,

$$dX(t) = \sigma \, dW(t),$$

where σ is a positive constant. As a result, an asset with initial price $X(0) = x$ would be worth

$$X(t) = x + \sigma W(t)$$

at time t. This approach was ahead of Bachelier's time, but it suffered from one serious flaw: for any $t > 0$ the price $X(t)$ can be negative with non-zero probability. Nevertheless, for short times it works well enough, since the probability is negligible. But as t increases, so does the probability that $X(t) < 0$, and the model departs from reality.

To remedy the flaw it was observed that investors work in terms of their potential gain or loss $dX(t)$ in proportion to the invested sum $X(t)$. Therefore, it is in fact the relative price $dX(t)/X(t)$ of an asset that reacts to the market fluctuations, i.e. should be proportional to $dW(t)$,

$$dX(t) = \sigma X(t) \, dW(t). \tag{7.1}$$

What is the precise mathematical meaning of this equality? Formally, it resembles a differential equation, but this immediately leads to a difficulty because the paths of $W(t)$ are nowhere differentiable. A way around the obstacle was found by Itô in the 1940s. In his hugely successful theory of *stochastic integrals* and *stochastic differential equations* Itô gave a rigorous meaning to equations such as (7.1) by writing them as integral equations involving a new kind of integral. In particular, (7.1) can be written as

$$X(t) = x + \sigma \int_0^t X(t)\, dW(t),$$

where the integral with respect to $W(t)$ on the right-hand side is called the *Itô stochastic integral* and will be defined in the next section. While at first sight one would expect the solution to this equation to be $xe^{W(t)}$, in fact it turns out to be

$$X(t) = xe^{W(t)}e^{-\frac{t}{2}},$$

which is the exponential martingale introduced in Exercise 6.35. The intriguing additional factor $e^{-\frac{t}{2}}$ is due to the non-differentiability of the paths of the Wiener process. Clearly, if $x > 0$, then $X(t) > 0$ for all $t \geq 0$, as required in the model of asset prices. In the following sections we shall learn how to transform and compute stochastic integrals and how to solve stochastic differential equations.

Throughout this chapter $W(t)$ will denote a Wiener process adapted to a filtration $\mathcal{F}_t$ and L^2 will be the space of square integrable random variables.

7.1 Itô Stochastic Integral: Definition

We shall follow a construction resembling that of the Riemann integral. First, the integral will be defined for a class of piecewise constant processes called random step processes. Then it will be extended to a larger class by approximation.

There are, however, at least two major differences between the Riemann and Itô integrals. One is the type of convergence. The approximations of the Riemann integral converge in $\mathbb{R}$, while the Itô integral will be approximated by sequences of random variables converging in L^2. The other difference is this. The Riemann sums approximating the integral of a function $f : [0,T] \to \mathbb{R}$ are of the form

$$\sum_{j=0}^{n-1} f(s_j)(t_{j+1} - t_j),$$

where $0 = t_0 < t_1 < \cdots < t_n = T$ and s_j is an *arbitrary* point in $[t_j, t_{j+1}]$ for each j. The value of the Riemann integral *does not* depend on the choice of the points $s_j \in [t_j, t_{j+1})$. In the stochastic case the approximating sums will have the form

$$\sum_{j=0}^{n-1} f(s_j)\left(W(t_{j+1}) - W(t_j)\right).$$

It turns out that the limit of such approximations *does* depend on the choice of the intermediate points s_j in $[t_j, t_{j+1}]$. In the next exercise we take $f(t) = W(t)$ and consider two different choices of intermediate points.

Exercise 7.1

Let $0 = t_0^n < t_1^n < \cdots < t_n^n = T$, where $t_j^n = \frac{jT}{n}$, be a partition of the interval $[0, T]$ into n equal parts. Find the following limits in L^2:

$$\lim_{n\to\infty} \sum_{j=0}^{n-1} W(t_j^n)\left(W(t_{j+1}^n) - W(t_j^n)\right)$$

and

$$\lim_{n\to\infty} \sum_{j=0}^{n-1} W(t_{j+1}^n)\left(W(t_{j+1}^n) - W(t_j^n)\right).$$

Hint Apply Exercise 6.29. You will need to transform the sums to make this possible. The identities

$$\begin{aligned} a(b-a) &= \frac{1}{2}\left(b^2 - a^2\right) - \frac{1}{2}(a-b)^2, \\ b(b-a) &= \frac{1}{2}\left(b^2 - a^2\right) + \frac{1}{2}(a-b)^2 \end{aligned}$$

may be of help.

The ambiguity resulting from different choices of the intermediate points s_j in each subinterval $[t_j, t_{j+1}]$ can be removed by insisting that the approximations should consist only of random variables adapted to the underlying filtration $\mathcal{F}_t$. This amounts to taking $s_j = t_j$ for each j. The choice is motivated by the interpretation of $\mathcal{F}_t$: the value of the approximation at t may depend only by what has happened up to time t, not on any future events.

Definition 7.1

We shall call $f(t), t \geq 0$ a *random step process* if there is a finite sequence of numbers $0 = t_0 < t_1 < \ldots < t_n$ and square integrable random variables

$\eta_0, \eta_1, \ldots, \eta_{n-1}$ such that

$$f(t) = \sum_{j=0}^{n-1} \eta_j 1_{[t_j, t_{j+1})}(t), \tag{7.2}$$

where η_j is $\mathcal{F}_{t_j}$-measurable for $j = 0, 1, \ldots, n-1$. The set of random step processes will be denoted by M^2_{step}.

Observe that the assumption that the η_j are to be $\mathcal{F}_{t_j}$-measurable ensures that $f(t)$ is adapted to the filtration $\mathcal{F}_t$. The assumption that the η_j are square integrable ensures that $f(t)$ is square integrable for each t. Also, M^2_{step} is a vector space, that is, $af + bg \in M^2_{\text{step}}$ for any $f, g \in M^2_{\text{step}}$ and $a, b \in \mathbb{R}$.

Definition 7.2

The *stochastic integral* of a random step process $f \in M^2_{\text{step}}$ of the form (7.2) is defined by

$$I(f) = \sum_{j=0}^{n-1} \eta_j \left(W(t_{j+1}) - W(t_j)\right). \tag{7.3}$$

Proposition 7.1

For any random step process $f \in M^2_{\text{step}}$ the stochastic integral $I(f)$ is a square integrable random variable, i.e. $I(f) \in L^2$, such that

$$E\left(|I(f)|^2\right) = E\left(\int_0^\infty |f(t)|^2 \, dt\right).$$

Proof

Let us denote the increment $W(t_{j+1}) - W(t_j)$ by $\Delta_j W$ and $t_{j+1} - t_j$ by $\Delta_j t$ for brevity. Then

$$E(\Delta_j W) = 0 \quad \text{and} \quad E\left(\Delta_j^2 W\right) = \Delta_j t.$$

First, we shall compute the expectation of

$$|I(f)|^2 = \sum_{j=0}^{n-1} \sum_{k=0}^{n-1} \eta_j \eta_k \Delta_j W \Delta_k W = \sum_{j=0}^{n-1} \eta_j^2 \Delta_j^2 W + 2 \sum_{k<j} \eta_j \eta_k \Delta_j W \Delta_k W.$$

Since η_j and $\Delta_j W$ are independent,

$$E\left(\eta_j^2 \Delta_j^2 W\right) = E\left(\eta_j^2\right) E\left(\Delta_j^2 W\right) = E\left(\eta_j^2\right) \Delta_j t.$$

If $k < j$, then $\eta_j \eta_k \Delta_k W$ and $\Delta_j W$ are independent, so

$$E(\eta_j \eta_k \Delta_j W \Delta_k W) = E(\eta_j \eta_k \Delta_k W) E(\Delta_j W) = 0.$$

Therefore

$$E\left(|I(f)|^2\right) = \sum_{j=0}^{n-1} E\left(\eta_j^2\right) \Delta_j t.$$

It follows that $I(f) \in L^2$, since $\eta_0, \eta_1, \ldots, \eta_{n-1} \in L^2$.

On the other hand,

$$|f(t)|^2 = \sum_{j=0}^{n-1} \sum_{k=0}^{n-1} \eta_j \eta_k 1_{[t_j, t_{j+1})}(t) 1_{[t_k, t_{k+1})}(t) = \sum_{j=0}^{n-1} \eta_j^2 1_{[t_j, t_{j+1})}(t),$$

implying that

$$E\left(\int_0^\infty |f(t)|^2 \, dt\right) = \sum_{j=0}^{n-1} E\left(\eta_j^2\right) \Delta_j t.$$

This means that

$$E\left(|I(f)|^2\right) = E\left(\int_0^\infty |f(t)|^2 \, dt\right),$$

as required. □

Exercise 7.2

Verify that for any random step processes $f, g \in M^2_{\text{step}}$

$$E(I(f)I(g)) = E\left(\int_0^\infty f(t)g(t) \, dt\right).$$

Hint Try to adapt the proof of Proposition 7.1. Use a common partition $0 = t_0 < t_1 < \cdots < t_n$ in which to represent both f and g in the form (7.2).

Exercise 7.3

Show that $I : M^2_{\text{step}} \to L^2$ is a linear map, i.e. for any $f, g \in M^2_{\text{step}}$ and any $\alpha, \beta \in \mathbb{R}$

$$I(\alpha f + \beta g) = \alpha I(f) + \beta I(g).$$

Hint As in Exercise 7.2, use a common partition $0 = t_0 < t_1 < \cdots < t_n$ in which to represent both f and g in the form (7.2).

The stochastic integral $I(f)$ has been defined for any random step process $f \in M^2_{\text{step}}$. The next stage is to extend I to a larger class of processes by approximation. This larger class can be defined as follows.

Definition 7.3

We denote by M^2 the class of stochastic processes $f(t), t \geq 0$ such that

$$E\left(\int_0^\infty |f(t)|^2 \, dt\right) < \infty$$

and there is a sequence $f_1, f_2, \ldots \in M^2_{\text{step}}$ of random step processes such that

$$\lim_{n\to\infty} E\left(\int_0^\infty |f(t) - f_n(t)|^2 \, dt\right) = 0. \tag{7.4}$$

In this case we shall say that the sequence of random step processes $f_1, f_2, \ldots$ *approximates* f in M^2.

Definition 7.4

We call $I(f) \in L^2$ the *Itô stochastic integral* (from 0 to ∞) of $f \in M^2$ if

$$\lim_{n\to\infty} E\left(|I(f) - I(f_n)|^2\right) = 0 \tag{7.5}$$

for any sequence $f_1, f_2, \ldots \in M^2_{\text{step}}$ of random step processes that approximates f in M^2, i.e. such that (7.4) is satisfied. We shall also write

$$\int_0^\infty f(t) \, dW(t)$$

in place of $I(f)$.

Proposition 7.2

For any $f \in M^2$ the stochastic integral $I(f) \in L^2$ exists, is unique (as an element of L^2, i.e. to within equality a.s.) and satisfies

$$E\left(|I(f)|^2\right) = E\left(\int_0^\infty |f(t)|^2 \, dt\right). \tag{7.6}$$

Proof

It will be convenient to write

$$\|f\|_{M^2} = \sqrt{E\left(\int_0^\infty |f(t)|^2 \, dt\right)} \quad \text{and} \quad \|\eta\|_{L^2} = \sqrt{E\left(\eta^2\right)}$$

for any $f \in M^2$ and $\eta \in L^2$. These are norms[1] in M^2 and L^2, respectively.

Let $f_1, f_2, \ldots \in M^2_{\text{step}}$ be a sequence of random step processes approximating $f \in M^2$, i.e. satisfying (7.4), which can be written as

$$\lim_{n\to\infty} \|f - f_n\|_{M^2} = 0.$$

We claim that $I(f_1), I(f_2), \ldots$ is a Cauchy sequence in L^2. Indeed, for any $\varepsilon > 0$ there is an N such that $\|f - f_n\|_{M^2} < \frac{\varepsilon}{2}$ for all $n > N$. By Proposition 7.1

$$\begin{aligned}
\|I(f_m) - I(f_n)\|_{L^2} &= \|I(f_m - f_n)\|_{L^2} \\
&= \|f_m - f_n\|_{M^2} \\
&\leq \|f - f_m\|_{M^2} + \|f - f_n\|_{M^2} \\
&< \frac{\varepsilon}{2} + \frac{\varepsilon}{2} = \varepsilon
\end{aligned}$$

for any $m, n > N$, which proves the claim.

Because L^2 with the norm $\|\cdot\|_{L^2}$ is a complete space (in fact a Hilbert space), every Cauchy sequence in L^2 has a limit. It follows that $I(f_1), I(f_2), \ldots$ has a limit in L^2 for any sequence $f_1, f_2, \ldots$ of random step processes approximating f. It remains to show that the limit is the same for all such sequences. Suppose that $f_1, f_2, \ldots$ and $g_1, g_2, \ldots$ are two sequences of random step processes approximating f. Then the interlaced sequence $f_1, g_1, f_2, g_2, \ldots$ approximates f too, so the sequence $I(f_1), I(g_1), I(f_2), I(g_2), \ldots$ has a limit in L^2. But then all subsequences of the latter sequence, in particular, $I(f_1), I(f_2), \ldots$ and $I(g_1), I(g_2), \ldots$ have the same limit, which we denote by $I(f)$. We have shown that

$$\lim_{n\to\infty} \|I(f) - I(f_n)\|_{L^2} = 0,$$

i.e. (7.5) holds for any sequence $f_1, f_2, \ldots$ of random step processes approximating f.

Finally, by Proposition 7.1

$$\|I(f_n)\|_{L^2} = \|f_n\|_{M^2}$$

for each n, since the f_n are random step processes. By taking the limit as $n \to \infty$ we obtain

$$\|I(f)\|_{L^2} = \|f\|_{M^2}.$$

But this is equality (7.6). □

[1] To be precise, the norms are defined on classes of functions, respectively, from M^2 and L^2 determined by the relation of equality a.s. However, we shall follow the custom of identifying such classes with any of their members.

Exercise 7.4

Show that for any $f, g \in M^2$

$$E\left(I(f)I(g)\right) = E\left(\int_0^\infty f(t)g(t)\,dt\right).$$

Hint Write the left-hand side in terms of $E\left(|I(f)+I(g)|^2\right)$ and $E\left(|I(f)-I(g)|^2\right)$, the right-hand side in terms of $E\left(\int_0^\infty |f(t)+g(t)|^2\,dt\right)$ and $E\left(\int_0^\infty |f(t)-g(t)|^2\,dt\right)$ and then use (7.6).

Having defined the Itô stochastic integral from 0 to ∞, we are now in a position to consider stochastic integrals over any finite time interval $[0, T]$.

Definition 7.5

For any $T > 0$ we shall denote by M_T^2 the space of all stochastic processes $f(t), t \geq 0$ such that

$$1_{[0,T)} f \in M^2$$

The Itô stochastic integral (from 0 to T) of $f \in M_T^2$ is defined by

$$I_T(f) = I\left(1_{[0,T)} f\right). \tag{7.7}$$

We shall also write

$$\int_0^T f(t)\,dW(t)$$

in place of $I_T(f)$.

Exercise 7.5

Show that each random step process $f \in M^2_{\text{step}}$ belongs to M_t^2 for any $t > 0$ and

$$I_t(f) = \int_0^t f(s)\,dW(s)$$

is a martingale.

Hint The stochastic integral of a random step process f is given by the sum (7.3). What is the conditional expectation of the jth term of this sum given $\mathcal{F}_s$ if $s < t_j$? What is it when $s \geq t_j$?

The processes for which the stochastic integral exists have been defined as those that can be approximated by random step processes. However, it is not always easy to check whether or not such an approximation exists. For practical purposes it is important to have a straightforward sufficient condition for a process to have a stochastic integral. In calculus there is a well-known result of this kind: the Riemann integral exists for any continuous function. Here is a theorem of this kind for the Itô integral.

Theorem 7.1

Let $f(t), t \geq 0$ be a stochastic process with a.s. continuous paths adapted to the filtration $\mathcal{F}_t$. Then

1) $f \in M^2$, i.e. the Itô integral $I(f)$ exists, whenever

$$E\left(\int_0^\infty |f(t)|^2 \, dt\right) < \infty; \tag{7.8}$$

2) $f \in M_T^2$, i.e. the Itô integral $I_T(f)$ exists, whenever

$$E\left(\int_0^T |f(t)|^2 \, dt\right) < \infty. \tag{7.9}$$

Proof

1) Suppose that $f(t), t \geq 0$ is an adapted process with a.s. continuous paths. If (7.8) holds, then

$$f_n(t) = \begin{cases} n \int_{\frac{k-1}{n}}^{\frac{k}{n}} f(s) \, ds & \frac{k}{n} \leq t < \frac{k+1}{n} \text{ for } k = 1, 2, \ldots, n^2 - 1, \\ 0 & \text{otherwise,} \end{cases} \tag{7.10}$$

is a sequence of random step processes in M^2_{step}. Observe that for any $k = 1, 2, \ldots$

$$\int_{\frac{k}{n}}^{\frac{k+1}{n}} |f_n(t)|^2 \, dt = n \left| \int_{\frac{k-1}{n}}^{\frac{k}{n}} f(t) dt \right|^2 \leq \int_{\frac{k-1}{n}}^{\frac{k}{n}} |f(t)|^2 \, dt \quad \text{a.s.} \tag{7.11}$$

by Jensen's inequality. We claim that

$$\lim_{n \to \infty} \int_0^\infty |f(t) - f_n(t)|^2 \, dt = 0 \quad \text{a.s.}$$

This will imply that

$$\lim_{n \to \infty} E\left(\int_0^\infty |f(t) - f_n(t)|^2 \, dt\right) = 0$$

by the dominated convergence theorem and condition (7.8) because

$$\begin{aligned} \int_0^\infty |f(t) - f_n(t)|^2 \, dt &\leq 2 \int_0^\infty \left(|f(t)|^2 + |f_n(t)|^2\right) dt \\ &\leq 4 \int_0^\infty |f(t)|^2 \, dt. \end{aligned}$$

The last inequality follows, since

$$\int_0^\infty |f_n(t)|^2 \, dt \leq \int_0^\infty |f(t)|^2 \, dt \quad \text{a.s.}$$

for any n, by taking the sum from $k = 0$ to ∞ in (7.11).

To verify the claim observe that

$$\begin{aligned}
\int_0^\infty |f(t) - f_n(t)|^2 \, dt &= \int_0^N |f(t) - f_n(t)|^2 \, dt + \int_N^\infty |f(t) - f_n(t)|^2 \, dt \\
&\leq \int_0^N |f(t) - f_n(t)|^2 \, dt + 2 \int_N^\infty \left(|f(t)|^2 + |f_n(t)|^2\right) dt \\
&\leq \int_0^N |f(t) - f_n(t)|^2 \, dt + 4 \int_{N-1}^\infty |f(t)|^2 \, dt \quad \text{a.s.}
\end{aligned}$$

The last inequality holds because

$$\int_N^\infty |f_n(t)|^2 \, dt \leq \int_{N-\frac{1}{n}}^\infty |f(t)|^2 \, dt \leq \int_{N-1}^\infty |f(t)|^2 \, dt \quad \text{a.s.}$$

for any n and N, by taking the sum from $k = nN$ to ∞ in (7.11). The claim follows because

$$\lim_{N\to\infty} \int_{N-1}^\infty |f(t)|^2 \, dt = 0 \quad \text{a.s.}$$

by (7.8) and

$$\lim_{n\to\infty} \int_0^N |f(t) - f_n(t)|^2 \, dt = 0 \quad \text{a.s.}$$

for any fixed N by the continuity of paths of f.

The above means that the sequence $f_1, f_2, \ldots \in M^2_{\text{step}}$ approximates f in the sense of Definition 7.3, so $f \in M^2$.

2) If f satisfies (7.9) for some $T > 0$, then $1_{[0,T)} f$ satisfies (7.8). Since f is adapted and has a.s. continuous paths, $1_{[0,T)} f$ is also adapted and its paths are a.s. continuous, except perhaps at T. But the lack of continuity at the single point T does not affect the argument in 1), so $1_{[0,T)} f \in M^2$. This in turn implies that $f \in M_T^2$, completing the proof. □

Exercise 7.6

Show that the Wiener process $W(t)$ belongs to M_T^2 for each $T > 0$.

Hint Apply part 2) of Theorem 7.1.

Exercise 7.7

Show that $W(t)^2$ belongs to M_T^2 for each $T > 0$.

Hint Once again, apply part 2) of Theorem 7.1.

The next theorem, which we shall state without proof, provides a characterization of M^2 and M_T^2, i.e. a necessary and sufficient condition for a stochastic process f to belong to M^2 or M_T^2. It involves the notion of a progressively measurable process.

Definition 7.6

A stochastic process $f(t), t \geq 0$ is called *progressively measurable* if for any $t \geq 0$

$$(s, \omega) \mapsto f(s, \omega)$$

is a measurable function form $[0, t] \times \Omega$ with the σ-field $\mathcal{B}[0, t] \bar{\times} \mathcal{F}$ to $\mathbb{R}$. Here $\mathcal{B}[0, t] \bar{\times} \mathcal{F}$ is the product σ-field on $[0, t] \times \Omega$, that is, the smallest σ-field containing all sets of the form $A \times B$, where $A \subset [0, t]$ is a Borel set and $B \in \mathcal{F}$.

Theorem 7.2

1) The space M^2 consists of all progressively measurable stochastic processes $f(t), t \geq 0$ such that

$$E\left(\int_0^\infty |f(t)|^2 \, dt\right) < \infty.$$

2) The space M_T^2 consists of all progressively measurable stochastic processes $f(t), t \geq 0$ such that

$$E\left(\int_0^T |f(t)|^2 \, dt\right) < \infty.$$

7.2 Examples

According to Exercise 7.4, the Wiener process $W(t)$ belongs to M_T^2 for any $T > 0$. Therefore the stochastic integral in the next exercise exists.

Exercise 7.8

Verify the equality

$$\int_0^T W(t) \, dW(t) = \frac{1}{2} W(T)^2 - \frac{1}{2} T$$

by computing the stochastic integral from the definition, that is, by approximating the integrand by random step functions.

Hint It is convenient to use a partition of the interval $[0,T]$ into n equal parts. The limit of the sums approximating the integral has been found in Exercise 7.1.

Exercise 7.9

Verify the equality

$$\int_0^T t\,dW(t) = TW(T) - \int_0^T W(t)\,dt,$$

by computing the stochastic integral from the definition. (The integral on the right-hand side is understood as a Riemann integral defined pathwise, i.e. separately for each $\omega \in \Omega$.)

Hint You may want to use the same partition of $[0,T]$ into n equal parts as in Solution 7.8. The sums approximating the stochastic integral can be transformed with the aid of the identity

$$c(b-a) = (db - ca) - b\,(d-c)\,.$$

Exercise 7.10

Show that $W(t)^2$ belongs to M_T^2 for each $T > 0$ and verify the equality

$$\int_0^T W(t)^2 dW(t) = \frac{1}{3}W(T)^3 - \int_0^T W(t)\,dt,$$

where the integral on the right-hand side is a Riemann integral.

Hint As in the exercises above, it is convenient to use the partition of $[0,T]$ into n equal parts. The identity

$$a^2\,(b-a) = \frac{1}{3}\left(b^3 - a^3\right) - a\,(b-a)^2 - \frac{1}{3}\,(b-a)^3$$

can be applied to transform the sums approximating the stochastic integral.

7.3 Properties of the Stochastic Integral

The basic properties of the Itô integral are summarized in the theorem below.

Theorem 7.3

The following properties hold for any $f, g \in M_t^2$, any $\alpha, \beta \in \mathbb{R}$, and any $0 \leq s < t$:

1) *linearity*

$$\int_0^t (\alpha f(r) + \beta g(r))\, dW(r) = \alpha \int_0^t f(r)\, dW(r) + \beta \int_0^t g(r)\, dW(r);$$

2) *isometry*

$$E\left(\left|\int_0^t f(r)\, dW(r)\right|^2\right) = E\left(\int_0^t |f(r)|^2\, dr\right);$$

3) *martingale property*

$$E\left(\int_0^t f(r)\, dW(r)\middle| \mathcal{F}_s\right) = \int_0^s f(r)\, dW(r).$$

Proof

1) If f and g belong to M_t^2, then $1_{[0,t)}f$ and $1_{[0,t)}g$ belong to M^2, so there are sequences $f_1, f_2, \ldots$ and $g_1, g_2, \ldots$ in M^2_{step} approximating $1_{[0,t)}f$ and $1_{[0,t)}g$. It follows that $1_{[0,t)}(\alpha f + \beta g)$ can be approximated by $\alpha f_1 + \beta g_1, \alpha f_2 + \beta g_2, \ldots$. By Exercise 7.3

$$I(\alpha f_n + \beta g_n) = \alpha I(f_n) + \beta I(g_n)$$

for each n. Taking the L^2 limit on both sides of this equality as $n \to \infty$, we obtain

$$I\left(1_{[0,t)}(\alpha f + \beta g)\right) = \alpha I(1_{[0,t)}f) + \beta I(1_{[0,t)}g),$$

which proves 1).

2) This follows by approximating $1_{[0,t)}f$ by random step processes in M^2_{step} and using Proposition 7.1.

3) If f belongs to M_t^2, then $1_{[0,t)}f$ belongs to M^2. Let $f_1, f_2, \ldots$ be a sequence of processes in M^2_{step} approximating $1_{[0,t)}f$. By Exercise 7.5

$$E\left(I\left(1_{[0,t)}f_n\right)|\mathcal{F}_s\right) = I\left(1_{[0,s)}f_n\right) \tag{7.12}$$

for each n. By taking the L^2 limit of both sides of this equality as $n \to \infty$, we shall show that

$$E\left(I\left(1_{[0,t)}f\right)|\mathcal{F}_s\right) = I\left(1_{[0,s)}f\right),$$

which is what needs to be proved. Indeed, observe that $1_{[0,s)}f_1, 1_{[0,s)}f_2, \ldots$ is a sequence in M^2_{step} approximating $1_{[0,s)}f$, so

$$I\left(1_{[0,s)}f_n\right) \to I\left(1_{[0,s)}f\right) \quad \text{in } L^2 \text{ as } n \to \infty.$$

Similarly, $1_{[0,t)}f_1, 1_{[0,t)}f_2, \ldots$ is also a sequence in M^2_{step} approximating $1_{[0,t)}f$, which implies that

$$I\left(1_{[0,t)}f_n\right) \to I\left(1_{[0,t)}f\right) \quad \text{in } L^2 \text{ as } n \to \infty.$$

The lemma below implies that

$$E\left(I\left(1_{[0,t)}f_n\right)|\mathcal{F}_s\right) \to E\left(I\left(1_{[0,t)}f\right)|\mathcal{F}_s\right) \quad \text{in } L^2 \text{ as } n \to \infty,$$

completing the proof. □

Lemma 7.1

If ξ and $\xi_1, \xi_2, \ldots$ are square integrable random variables such that $\xi_n \to \xi$ in L^2 as $n \to \infty$, then

$$E(\xi_n|\mathcal{G}) \to E(\xi|\mathcal{G}) \quad \text{in } L^2 \text{ as } n \to \infty$$

for any σ-field $\mathcal{G}$ on Ω contained in $\mathcal{F}$.

Proof

By Jensen's inequality, see Theorem 2.2,

$$|E(\xi_n|\mathcal{G}) - E(\xi|\mathcal{G})|^2 = |E(\xi_n - \xi|\mathcal{G})|^2 \leq E\left(|\xi_n - \xi|^2\middle|\mathcal{G}\right),$$

which implies that

$$\begin{aligned} E\left(|E(\xi_n|\mathcal{G}) - E(\xi|\mathcal{G})|^2\right) &\leq E\left(E\left(|\xi_n - \xi|^2\middle|\mathcal{G}\right)\right) \\ &= E\left(|\xi_n - \xi|^2\right) \to 0 \end{aligned}$$

as $n \to \infty$. □

In the next theorem we consider the stochastic integral $\int_0^t f(s)\,dW(s)$ as a function of the upper integration limit t. Similarly as for the Riemann integral, it is natural to ask if this is a continuous function of t. The answer to this question involves the notion of a modification of a stochastic process.

Definition 7.7

Let $\xi(t)$ and $\zeta(t)$ be stochastic processes defined for $t \in T$, where $T \subset \mathbb{R}$. We say that the processes are *modifications* (or *versions*) of one another if

$$P\{\xi(t) = \zeta(t)\} = 1 \quad \text{for all } t \in T. \tag{7.13}$$

Remark 7.1

If $T \subset \mathbb{R}$ is a countable set, then (7.13) is equivalent to the condition

$$P\{\xi(t) = \zeta(t) \text{ for all } t \in T\} = 1.$$

However, this is not necessarily so if T is uncountable.

The following result is stated without proof.

Theorem 7.4

Let $f(s)$ be a process belonging to M_t^2 and let

$$\xi(t) = \int_0^t f(s)\,dW(s)$$

for every $t \geq 0$. Then there exists an adapted modification $\zeta(t)$ of $\xi(t)$ with a.s. continuous paths. This modification is unique up to equality a.s.

From now on we shall always identify $\int_0^t f(s)\,dW(s)$ with the adapted modification having a.s. continuous paths. This convention works beautifully together with Theorem 7.1 whenever there is a need to show that a stochastic integral can be used as the integrand of another stochastic integral, i.e. belongs to M_T^2 for $T \geq 0$. This is illustrated by the next exercise.

Exercise 7.11

Show that

$$\xi(t) = \int_0^t W(s)\,dW(s)$$

belongs to M_T^2 for any $T \geq 0$.

Hint By Theorem 7.4 $\xi(t)$ can be identified with an adapted modification having a.s. continuous trajectories. Because of this, it suffices to verify that $\xi(t)$ satisfies condition (7.9) of Theorem 7.1.

7.4 Stochastic Differential and Itô Formula

Any continuously differentiable function $x(t)$ such that $x(0) = 0$ satisfies the formulae

$$x(T)^2 = 2\int_0^T x(t)\,dx(t),$$
$$x(T)^3 = 3\int_0^T x(t)^2\,dx(t),$$

where $dx(t)$ can simply be understood as a shorthand notation for $x'(t)\,dt$, the integrals on the right-hand side being Riemann integrals. Similar formulae have been obtained in Exercises 7.8 and 7.10 for the Wiener process:

$$W(T)^2 = \int_0^T dt + 2\int_0^T W(t)\,dW(t),$$
$$W(T)^3 = 3\int_0^T W(t)\,dt + 3\int_0^T W(t)^2 dW(t).$$

Here the stochastic integrals resemble the corresponding expressions for a smooth function $x(t)$, but there are also the intriguing terms $\int_0^T dt$ and $3\int_0^T W(t)\,dt$. The formulae for $W(T)^2$ and $W(T)^3$ are examples of the much more general *Itô formula*, a crucial tool for transforming and computing stochastic integrals. Terms such as $\int_0^T dt$ and $3\int_0^T W(t)\,dt$, which have no analogues in the classical calculus of smooth functions, are a feature inherent in the Itô formula and referred to as the *Itô correction*. The class of processes appearing in the Itô formula is defined as follows.

Definition 7.8

A stochastic process $\xi(t), t \geq 0$ is called an *Itô process* if it has a.s. continuous paths and can be represented as

$$\xi(T) = \xi(0) + \int_0^T a(t)\,dt + \int_0^T b(t)\,dW(t) \quad \text{a.s.,} \tag{7.14}$$

where $b(t)$ is a process belonging to M_T^2 for all $T > 0$ and $a(t)$ is a process adapted to the filtration $\mathcal{F}_t$ such that

$$\int_0^T |a(t)|\,dt < \infty \quad \text{a.s.} \tag{7.15}$$

for all $T \geq 0$. The class of all adapted processes $a(t)$ satisfying (7.15) for some $T > 0$ will be denoted by $\mathcal{L}_T^1$.

For an Itô process ξ it is customary to write (7.14) as

$$d\xi(t) = a(t)\,dt + b(t)\,dW(t) \tag{7.16}$$

and to call $d\xi(t)$ the *stochastic differential* of $\xi(t)$. This is known as the *Itô differential notation*. It should be emphasized that the stochastic differential has no well-defined mathematical meaning on its own and should always be understood in the context of the rigorous equation (7.14). The Itô differential notation is an efficient way of writing this equation, rather than an attempt to give a precise mathematical meaning to the stochastic differential.

Example 7.1

The Wiener process $W(t)$ satisfies

$$W(T) = \int_0^T dW(t).$$

(The right-hand side is the stochastic integral $I(f)$ of the random step process $f = 1_{[0,T)}$.) This is an equation of the form (7.14) with $a(t) = 0$ and $b(t) = 1$, which belong, respectively, to $\mathcal{L}_T^1$ and $\mathcal{M}_T^2$ for any $T \geq 0$. It follows that the Wiener process is an Itô process.

Example 7.2

Every process of the form

$$\xi(T) = \xi(0) + \int_0^T a(t)\, dt,$$

where $a(t)$ is a process belonging to $\mathcal{L}_T^1$ for any $T \geq 0$, is an Itô process. In particular, every deterministic process of this form, where $a(t)$ is a deterministic integrable function, is an Itô process.

Example 7.3

Since $a(t) = 1$ and $b(t) = 2W(t)$ belong, respectively, to the classes $\mathcal{L}_T^1$ and $\mathcal{M}_T^2$ for each $T \geq 0$,

$$W(T)^2 = \int_0^T dt + 2\int_0^T W(t)\, dW(t)$$

is an Itô process; see Exercise 7.8. The last equation can also be written as

$$d\left(W(t)^2\right) = dt + 2\, W(t)\, dW(t),$$

providing a formula for the stochastic differential $d\left(W(t)^2\right)$ of $W(t)^2$.

Exercise 7.12

Show that $W(t)^3$ is an Itô process and find a formula for the stochastic differential $d\left(W(t)^3\right)$.

Hint Refer to Exercise 7.10.

Exercise 7.13

Show that $tW(t)$ is an Itô process and find a formula for the stochastic differential $d(tW(t))$.

Hint Use Exercise 7.9.

The above examples and exercises are particular cases of an extremely important general formula for transforming stochastic differentials established by Itô. To begin with, we shall state and prove a simplified version of the formula, followed by the general theorem. The proof of the simplified version captures the essential ingredients of the somewhat tedious general argument, which will be omitted. In fact, many of the essential ingredients of the proof are already present in the examples and exercises considered above.

Theorem 7.5 (Itô formula, simplified version)

Suppose that $F(t,x)$ is a real-valued function with continuous partial derivatives $F'_t(t,x)$, $F'_x(t,x)$ and $F''_{xx}(t,x)$ for all $t \geq 0$ and $x \in \mathbb{R}$. We also assume that the process $F'_x(t,W(t))$ belongs to M^2_T for all $T \geq 0$. Then $F(t,W(t))$ is an Itô process such that

$$F(T,W(T)) - F(0,W(0)) = \int_0^T \left(F'_t(t,W(t)) + \frac{1}{2}F''_{xx}(t,W(t))\right) dt + \int_0^T F'_x(t,W(t))\, dW(t) \quad \text{a.s.} \tag{7.17}$$

In differential notation this formula can be written as

$$dF(t,W(t)) = \left(F'_t(t,W(t)) + \frac{1}{2}F''_{xx}(t,W(t))\right) dt + F'_x(t,W(t))\, dW(t). \tag{7.18}$$

Remark 7.2

Compare the latter with the chain rule

$$dF(t,x(t)) = F'_t(t,x(t))\, dt + F'_x(t,x(t))\, dx(t).$$

for a smooth function $x(t)$, where $dx(t)$ is understood as a shorthand notation for $x'(t)\, dt$. The additional term $\frac{1}{2}F''_{xx}(t,W(t))\, dt$ in (7.18) is called the *Itô correction.*

Proof

First we shall prove the Itô formula under the assumption that F and the partial derivatives F'_x and F''_{xx} are bounded by some $C > 0$.

Consider a partition $0 = t_0^n < t_1^n < \cdots < t_n^n = T$, where $t_i^n = \frac{iT}{n}$, of $[0,T]$ into n equal parts. We shall denote the increments $W(t_{i+1}^n) - W(t_i^n)$ by $\Delta_i^n W$ and $t_{i+1}^n - t_i^n$ by $\Delta_i^n t$. We shall also write W_i^n instead of $W(t_i^n)$ for brevity. According to the Taylor formula, there is a point $\tilde{W}_i^n$ in each interval $[W(t_i^n), W(t_{i+1}^n)]$ and a point $\tilde{t}_i^n$ in each interval $[t_i^n, t_{i+1}^n]$ such that

$$F(T,W(T)) - F(0,W(0)) = \sum_{i=0}^{n-1} \left(F(t_{i+1}^n, W_{i+1}^n) - F(t_i^n, W_i^n)\right)$$

$$= \sum_{i=0}^{n-1} \left(F(t_{i+1}^n, W_{i+1}^n) - F(t_i^n, W_{i+1}^n)\right) + \sum_{i=0}^{n-1} \left(F(t_i^n, W_{i+1}^n) - F(t_i^n, W_i^n)\right)$$

$$= \sum_{i=0}^{n-1} F_t'(\tilde{t}_i^n, W_{i+1}^n)\Delta_i^n t + \sum_{i=0}^{n-1} F_x'(t_i^n, W_i^n)\Delta_i^n W + \frac{1}{2}\sum_{i=0}^{n-1} F_{xx}''(t_i^n, \tilde{W}_i^n)\left(\Delta_i^n W\right)^2$$

$$= \sum_{i=0}^{n-1} F_t'(\tilde{t}_i^n, W_{i+1}^n)\Delta_i^n t + \frac{1}{2}\sum_{i=0}^{n-1} F_{xx}''(t_i^n, W_i^n)\Delta_i^n t + \sum_{i=0}^{n-1} F_x'(t_i^n, W_i^n)\Delta_i^n W$$

$$+ \frac{1}{2}\sum_{i=0}^{n-1} F_{xx}''(t_i^n, W_i^n)\left(\left(\Delta_i^n W\right)^2 - \Delta_i^n t\right)$$

$$+ \frac{1}{2}\sum_{i=0}^{n-1} \left[F_{xx}''(t_i^n, \tilde{W}_i^n) - F_{xx}''(t_i^n, W_i^n)\right]\left(\Delta_i^n W\right)^2.$$

We shall deal separately with each sum in the last expression, splitting the proof into several steps.

Step 1. We claim that

$$\lim_{n\to\infty} \sum_{i=0}^{n-1} F_t'(\tilde{t}_i^n, W_{i+1}^n)\Delta_i^n t = \int_0^T F_t'(t, W(t))\, dt \quad \text{a.s.}$$

This is because the paths of $W(t)$ are a.s. continuous, and $F_t'(t,x)$ is continuous as a function of two variables by assumption. Indeed, every continuous path of the Wiener process is bounded on $[0,T]$, i.e. there is an $M > 0$, which may depend on the path, such that

$$|W(t)| \le M \quad \text{for all } t \in [0,T].$$

As a continuous function, $F_t'(t,x)$ is uniformly continuous on the compact set $[0,T] \times [-M,M]$ and W is uniformly continuous on $[0,T]$. It follows that

$$\lim_{n\to\infty} \sup_{i,t} \left|F_t'(\tilde{t}_i^n, W_{i+1}^n) - F_t'(t, W(t))\right| = 0 \quad \text{a.s.},$$

where the supremum is taken over all $i = 0, \ldots, n-1$ and $t \in [t_i^n, t_{i+1}^n]$. By the definition of the Riemann integral this proves the claim.

Step 2. This is very similar to Step 1. By continuity

$$\lim_{n\to\infty} \sup_{i,t} |F''_{xx}(t_i^n, W_i^n) - F''_{xx}(t, W(t))| = 0 \quad \text{a.s.},$$

where the supremum is taken over all $i = 0, \dots, n-1$ and $t \in [t_i^n, t_{i+1}^n]$. By the definition of the Riemann integral

$$\lim_{n\to\infty} \sum_{i=0}^{n-1} F''_{xx}(t_i^n, W_i^n)\Delta_i^n t = \int_0^T F''_{xx}(t, W(t))\, dt \quad \text{a.s.}$$

Step 3. We shall verify that

$$\lim_{n\to\infty} \sum_{i=0}^{n-1} F'_x(t_i^n, W_i^n)\Delta_i^n W = \int_0^T F'_x(t, W(t))\, dW(t) \quad \text{in } L^2.$$

If $F'_x(t, x)$ is bounded by $C > 0$, then $f(t) = F'_x(t, W(t))$ belongs to M_T^2 by Theorem 7.1, and the sequence of random step processes

$$f_n = \sum_{i=0}^{n-1} F'_x(t_i^n, W_i^n) 1_{[t_i^n, t_{i+1}^n)} \in M_{\text{step}}^2$$

approximates f. Indeed, by continuity

$$\lim_{n\to\infty} |f_n(t) - f(t)|^2 = 0 \quad \text{for any } t \in [0, T], \quad \text{a.s.}$$

Because $|f_n(t) - f(t)|^2 \leq 2C^2$, it follows that

$$\lim_{n\to\infty} \int_0^T |f_n(t) - f(t)|^2\, dt = 0 \quad \text{a.s.}$$

by Lebesgue's dominated convergence theorem. But $\int_0^T |f_n(t) - f(t)|^2\, dt \leq 2TC^2$, so

$$\lim_{n\to\infty} E\left(\int_0^T |f_n(t) - f(t)|^2\, dt\right) = 0$$

again by Lebesgue's dominated convergence theorem. This shows that f_n approximates f, which in turn implies that $I(f_n)$ tends to $I(f)$ in L^2, concluding Step 3.

Step 4. If F''_{xx} is bounded by $C > 0$, then

$$\lim_{n\to\infty} \sum_{i=0}^{n-1} F''_{xx}(t_i^n, W_i^n)\left((\Delta_i^n W)^2 - \Delta_i^n t\right) = 0 \quad \text{in } L^2,$$

since

$$
\begin{aligned}
E & \left| \sum_{i=0}^{n-1} F''_{xx}(t_i^n, W_i^n) \left((\Delta_i^n W)^2 - \Delta_i^n t \right) \right|^2 \\
&= \sum_{i=0}^{n-1} E \left| F''_{xx}(t_i^n, W_i^n) \left((\Delta_i^n W)^2 - \Delta_i^n t \right) \right|^2 \\
&= \sum_{i=0}^{n-1} E \left| F''_{xx}(t_i^n, W_i^n) \right|^2 E \left| (\Delta_i^n W)^2 - \Delta_i^n t \right|^2 \\
&\leq C^2 \sum_{i=0}^{n-1} E \left| (\Delta_i^n W)^2 - \Delta_i^n t \right|^2 = 2C^2 \sum_{i=0}^{n-1} (\Delta_i^n t)^2 \\
&= 2C^2 \sum_{i=0}^{n-1} \frac{T^2}{n^2} = 2C^2 \frac{T^2}{n} \to 0 \quad \text{as } n \to \infty.
\end{aligned}
$$

The first equality above holds because for any $i < j$

$$
\begin{aligned}
E & \left[F''_{xx}(t_i^n, W_i^n) \left((\Delta_i^n W)^2 - \Delta_i^n t \right) F''_{xx}(t_j^n, W_j^n) \left((\Delta_j^n W)^2 - \Delta_j^n t \right) \right] \\
&= E \left[F''_{xx}(t_i^n, W_i^n) \left((\Delta_i^n W)^2 - \Delta_i^n t \right) F''_{xx}(t_j^n, W_j^n) \right] E \left[(\Delta_j^n W)^2 - \Delta_j^n t \right] \\
&= 0.
\end{aligned}
$$

This is because the expressions in the last two square brackets are independent and the last expectation is equal to zero.

Step 5. By a similar continuity argument as in Steps 1 and 2

$$
\lim_{n \to \infty} \sup_i \left| F''_{xx}(t_i^n, \tilde{W}_i^n) - F''_{xx}(t_i^n, W_i^n) \right| = 0 \quad \text{a.s.},
$$

where the supremum is taken over all $i = 0, 1, \ldots, n-1$. Since $\sum_{i=0}^{n-1} (\Delta_i^n W)^2 \to t$ in L^2 as $n \to \infty$, there is a subsequence $n_1 < n_2 < \ldots$ such that

$$
\sum_{i=0}^{n_k - 1} (\Delta_i^{n_k} W)^2 \to t \quad \text{a.s.}
$$

as $k \to \infty$. It follows that

$$
\begin{aligned}
& \left| \sum_{i=0}^{n_k - 1} \left(F''_{xx}(t_i^{n_k}, \tilde{W}_i^{n_k}) - F''_{xx}(t_i^{n_k}, W_i^{n_k}) \right) (\Delta_i^{n_k} W)^2 \right| \\
& \qquad \leq \sup_i \left| F''_{xx}(t_i^{n_k}, \tilde{W}_{i+1}^{n_k}) - F''_{xx}(t_i^{n_k}, W_{i+1}^{n_k}) \right| \sum_{i=0}^{n_k - 1} (\Delta_i^{n_k} W)^2 \to 0 \quad \text{a.s.}
\end{aligned}
$$

as $k \to \infty$.

In those steps above where L^2 convergence was obtained, we also have convergence a.s. by taking a subsequence. This proves the Itô formula (7.17) under the assumption that the partial derivatives $F'_x(t,x)$ and $F''_{xx}(t,x)$ are bounded. To complete the proof we need to remove this assumption. Let $F(t,x)$ be an arbitrary function satisfying the conditions of Theorem 7.5. For each positive integer n take a smooth function φ_n from $\mathbb{R}$ to $[0,1]$ such that $\varphi_n(x) = 1$ for any $x \in [-n,n]$ and $\varphi_n(x) = 0$ for any $x \in [-n-1,n+1]$. Then

$$F_n(t,x) = \varphi_n(x)F(t,x)$$

also satisfies the conditions of Theorem 7.5 and has bounded partial derivatives $(F_n)'_x(t,x)$ and $(F_n)''_{xx}(t,x)$ for each n. Therefore, by the first part of the proof

$$\begin{aligned} &F_n(T,W(T)) - F_n(0,W(0)) \\ &= \int_0^T \left((F_n)'_t(t,W(t)) + \frac{1}{2}(F_n)''_{xx}(t,W(t)) \right) dt + \int_0^T (F_n)'_x(t,W(t))\,dW(t). \end{aligned}$$

Consider the expanding sequence of events

$$A_n = \left\{ \sup_{t\in[0,T]} |W(t)| < n \right\}.$$

Since $F(t,x) = F_n(t,x)$ for every $t \in [0,T]$ and $x \in [-n,n]$, it follows that (7.17) holds on A_n. It remains to show that

$$\lim_{n\to\infty} P(A_n) = 1$$

to prove that (7.17) holds a.s. But the latter is true because of Doob's maximal L^2 inequality, Theorem 6.7, which implies that

$$\begin{aligned} n^2(1 - P(A_n)) &= n^2 P\left\{ \sup_{t\in[0,T]} |W(t)| \geq n \right\} \\ &\leq E\left(\sup_{t\in[0,T]} |W(t)| \right)^2 \\ &\leq 4E|W(T)|^2 = 4T, \end{aligned}$$

completing the proof. □

Example 7.4

For $F(t,x) = x^2$ we have $F'_t(t,x) = 0$, $F'_x(t,x) = 2x$ and $F''_{xx}(t,x) = 2$. The Itô formula gives

$$d\left(W(t)^2\right) = dt + 2W(t)\,dW(t),$$

which is the same equality as in Exercise 7.8.

Example 7.5

For $F(t,x) = x^3$ we have $F'_t(t,x) = 0$, $F'_x(t,x) = 3x^2$ and $F''_{xx}(t,x) = 6x$. By the Itô formula we obtain the same equality

$$d\left(W(t)^3\right) = 3W(t)\,dt + 3W(t)^2\,dW(t)$$

as in Exercise 7.10.

Exercise 7.14 (exponential martingale)

Show that the exponential martingale $X(t) = e^{W(t)}e^{-\frac{t}{2}}$ is an Itô process and verify that it satisfies the equation

$$dX(t) = X(t)\,dW(t).$$

Hint Use the Itô formula with $F(t,x) = e^x e^{-\frac{t}{2}}$.

As compared with the simplified version just proved, in the general Itô formula below $W(t)$ will replaced by an arbitrary Itô process $\xi(t)$ such that

$$d\xi(t) = a(t)\,dt + b(t)\,dW(t), \tag{7.19}$$

where a belongs to $\mathcal{M}_t^2$ and b to $\mathcal{L}_t^1$ for all $t \geq 0$. In the general case the proof will be omitted.

Theorem 7.6 (Itô formula, general case)

Let $\xi(t)$ be an Itô process as above. Suppose that $F(t,x)$ is a real-valued function with continuous partial derivatives $F'_t(t,x)$, $F'_x(t,x)$ and $F''_{xx}(t,x)$ for all $t \geq 0$ and $x \in \mathbb{R}$. We also assume that the process $b(t)F'_x(t,\xi(t))$ belongs to $\mathcal{M}_T^2$ for all $T \geq 0$. Then $F(t,\xi(t))$ is an Itô process such that

$$\begin{aligned} dF(t,\xi(t)) = {} & \left(F'_t(t,\xi(t)) + F'_x(t,\xi(t))\,a(t) + \frac{1}{2}F''_{xx}(t,\xi(t))\,b(t)^2\right)dt \\ & + F'_x(t,\xi(t))\,b(t)\,dW(t). \end{aligned} \tag{7.20}$$

A convenient way to remember the Itô formula is to write down the Taylor expansion for $F(t,x)$ up to the terms with partial derivatives of order two, substituting $\xi(t)$ for x and the expression on the right-hand side of (7.19) for $d\xi(t)$, and using the so-called *Itô multiplication table*

$$\begin{aligned} dt\,dt &= 0, & dt\,dW(t) &= 0, \\ dW(t)\,dt &= 0, & dW(t)\,dW(t) &= dt. \end{aligned}$$

This informal procedure gives

$$\begin{aligned} dF &= F_t' \, dt + F_x' \, d\xi + \frac{1}{2} F_{tt}'' \, dt \, dt + F_{tx}'' \, dt \, d\xi + \frac{1}{2} F_{xx}'' \, d\xi \, d\xi \\ &= F_t' \, dt + F_x' \, (a \, dt + b \, dW) \\ &\quad + \frac{1}{2} F_{tt}'' \, dt \, dt + F_{tx}'' \, dt \, (a \, dt + b \, dW) + \frac{1}{2} F_{xx}'' \, (a \, dt + b \, dW) \, (a \, dt + b \, dW) \\ &= F_t' \, dt + F_x' \, (a \, dt + b \, dW) + \frac{1}{2} F_{xx}'' b^2 \, dt \\ &= \left(F_t' + F_x' a + \frac{1}{2} F_{xx}'' b^2 \right) dt + F_x' b \, dW, \end{aligned}$$

which is the expression in (7.20). Here we have omitted the arguments $(t, \xi(t))$ and, respectively, (t) in all functions for brevity.

Exercise 7.15

Applying the Itô formula to $F(t, x) = x^n$, show that

$$d \, W(t)^n = \frac{n(n-1)}{2} W(t)^{n-2} \, dt + n W(t)^{n-1} \, dW(t) \tag{7.21}$$

Hint This is a direct application of the Itô formula, but be careful with the assumptions, in particular make sure that $nW(t)^{n-1}$ belongs to M_T^2 for all $T > 0$.

Exercise 7.16 (Ornstein–Uhlenbeck process)

Suppose that $\alpha > 0$ and $\sigma \in \mathbb{R}$ are fixed. Define $Y(t), t \geq 0$ to be an adapted modification of the Itô integral

$$Y(t) = \sigma e^{-\alpha t} \int_0^t e^{\alpha s} \, dW(s)$$

with a.s. continuous paths. Show that $Y(t)$ satisfies

$$dY(t) = -\alpha Y(t) \, dt + \sigma \, dW(t)$$

Hint $Y(t) = F(t, \xi(t))$ with $\xi(t) = \sigma \int_0^t e^{\alpha s} \, dW(s)$ and $F(t, x) = e^{-\alpha t} x$.

7.5 Stochastic Differential Equations

This section will be devoted to *stochastic differential equations* of the form

$$d\xi(t) = f(\xi(t)) \, dt + g(\xi(t)) \, dW(t).$$

Solutions will be sought in the class of Itô processes $\xi(t)$ with a.s. continuous paths. As in the theory of ordinary differential equations, we need to specify an initial condition

$$\xi(0) = \xi_0.$$

Here ξ_0 can be a fixed real number or, in general, a random variable. Being an Itô process, $\xi(t)$ must be adapted to the filtration $\mathcal{F}_t$ of $W(t)$, so ξ_0 must be $\mathcal{F}_0$-measurable.

Example 7.6

The stochastic differential equation

$$dX(t) = X(t)\, dW(t) \tag{7.22}$$

was used as a motivation for developing Itô stochastic calculus at the beginning of the present chapter. In Exercise 7.14 it was verified that the exponential martingale

$$X(t) = e^{W(t)} e^{-\frac{t}{2}}$$

satisfies (7.22). It also satisfies the initial condition $X(0) = 1$. This is an example of a linear stochastic differential equation. For the solution of a general equation of this type with an arbitrary initial condition, see Exercise 7.20.

Example 7.7

In Exercise 7.16 it was shown that the Ornstein–Uhlenbeck process

$$Y(t) = \sigma e^{-\alpha t} \int_0^t e^{\alpha s}\, dW(s)$$

satisfies the stochastic differential equation

$$dY(t) = -\alpha Y(t)\, dt + \sigma\, dW(t)$$

with initial condition $Y(0) = 0$. This is an example of an inhomogeneous linear stochastic differential equation. See Exercise 7.17 for a solution with an arbitrary initial condition.

Definition 7.9

An Itô process $\xi(t)$, $t \geq 0$ is called a solution of the initial value problem

$$\begin{aligned} d\xi(t) &= f(\xi(t))\, dt + g(\xi(t))\, dW(t), \\ \xi(0) &= \xi_0 \end{aligned}$$

if ξ_0 is an $\mathcal{F}_0$-measurable random variable, the processes $f(\xi(t))$ and $g(\xi(t))$ belong, respectively, to $\mathcal{L}_T^1$ and M_T^2, and

$$\xi(T) = \xi_0 + \int_0^T f(\xi(t))\,dt + \int_0^T g(\xi(t))\,dW(t) \quad \text{a.s.} \tag{7.23}$$

for all $T \geq 0$.

Remark 7.3

In view of this definition, the notion of a stochastic differential equation is a fiction. In fact, only *stochastic integral equations* of the form (7.23) have a rigorous mathematical meaning. However, it proves convenient to use stochastic differentials informally and talk of stochastic differential equations to draw on the analogy with ordinary differential equations. This analogy will be employed to solve some stochastic differential equations later on in this section.

The existence and uniqueness theorem below resembles that in the theory of ordinary differential equations, where it is also crucial for the right-hand side of the equation to be Lipschitz continuous as a function of the solution.

Theorem 7.7

Suppose that f and g are Lipschitz continuous functions form $\mathbb{R}$ to $\mathbb{R}$, i.e. there is a constant $C > 0$ such that for any $x, y \in \mathbb{R}$

$$\begin{aligned} |f(x) - f(y)| &\leq C\,|x - y|\,, \\ |g(x) - g(y)| &\leq C\,|x - y|\,. \end{aligned}$$

Moreover, let ξ_0 be an $\mathcal{F}_0$-measurable square integrable random variable. Then the initial value problem

$$d\xi(t) = f(\xi(t))\,dt + g(\xi(t))\,dW(t), \tag{7.24}$$

$$\xi(0) = \xi_0 \tag{7.25}$$

has a solution $\xi(t), t \geq 0$ in the class of Itô processes. The solution is unique in the sense that if $\eta(t), t \geq 0$ is another Itô process satisfying (7.24) and (7.25), then the two processes are identical a.s., that is,

$$P\{\xi(t) = \eta(t) \text{ for all } t \geq 0\} = 1.$$

Proof (outline)

Let us fix $T > 0$. We are looking for a process $\xi \in M_T^2$ such that

$$\xi(s) = \xi_0 + \int_0^s f(\xi(t))\,dt + \int_0^s g(\xi(t))\,dW(t) \quad \text{a.s.} \tag{7.26}$$

for all $s \in [0, T]$. To obtain a solution to the stochastic differential equation (7.24) with initial condition (7.25) it suffices to take a modification of ξ with a.s. continuous paths, which exists by Theorem 7.4.

To show that a solution to the stochastic integral equation (7.26) exists we shall employ the Banach fixed point theorem in M_T^2 with the norm

$$\|\xi\|_\lambda^2 = E\int_0^T e^{-\lambda t}|\xi(t)|^2\,dt, \tag{7.27}$$

which turns M_T^2 into a complete normed vector space. The number $\lambda > 0$ should be chosen large enough, see below. To apply the fixed point theorem define $\Phi : M_T^2 \to M_T^2$ by

$$\Phi(\xi)(s) = \xi_0 + \int_0^s f(\xi(t))\,dt + \int_0^s g(\xi(t))\,dW(t) \tag{7.28}$$

for any $\xi \in M_T^2$ and $s \in [0, T]$. We claim that Φ is a strict contraction, i.e.

$$\|\Phi(\xi) - \Phi(\zeta)\|_\lambda \le \alpha\|\xi - \zeta\|_\lambda \tag{7.29}$$

for some $\alpha < 1$ and all $\xi, \zeta \in M_T^2$. Then, by the Banach theorem, Φ has a unique fixed point $\xi = \Phi(\xi)$. This is the desired solution to (7.26).

It remains to verify that Φ is indeed a strict contraction. It suffices to show that the two maps Φ_1 and Φ_2, where

$$\Phi_1(\xi)(s) = \int_0^s f(\xi(t))\,dt, \qquad \Phi_2(\xi)(s) = \int_0^s g(\xi(t))\,dW(t),$$

are strict contractions. For Φ_1 this follows from the Lipschitz continuity of f. For Φ_2 we need to use the Lipschitz continuity of g and the isometry property of the Itô integral. Let us mention just one essential step in the latter case. For any $\xi, \zeta \in M_T^2$

$$\begin{aligned}
\|\Phi_2(\xi) - \Phi_2(\zeta)\|_\lambda^2 &= E\int_0^T e^{-\lambda s}\left|\int_0^s [g(\xi(t)) - g(\zeta(t))]\,dW(t)\right|^2 ds \\
&= E\int_0^T e^{-\lambda s}\int_0^s |g(\xi(t)) - g(\zeta(t))|^2\,dt\,ds \\
&\le C^2 E\int_0^T e^{-\lambda s}\int_0^s |\xi(t) - \zeta(t)|^2\,dt\,ds
\end{aligned}$$

$$= C^2 E \int_0^T \left(\int_t^T e^{-\lambda s} e^{\lambda t}\, ds \right) e^{-\lambda t} |\xi(t) - \zeta(t)|^2 \, dt$$

$$\leq \frac{C^2}{\lambda} E \int_0^T e^{-\lambda t} |\xi(t) - \zeta(t)|^2 \, dt = \frac{C^2}{\lambda} \|\xi - \zeta\|_\lambda ,$$

since $\int_t^T e^{-\lambda s} e^{\lambda t}\, ds = \frac{1}{\lambda}(1 - e^{-\lambda(T-t)}) \leq \frac{1}{\lambda}$. Here C is the Lipschitz constant of g. If $\lambda > C^2$, then Φ_2 is a strict contraction.

There remain some technical points to be settled, but the main idea of the proof is shown above. □

Exercise 7.17

Find a solution of the stochastic differential equation

$$dX(t) = -\alpha X(t)\, dt + \sigma\, dW(t)$$

with initial condition $X(0) = x_0$, where x_0 is an arbitrary real number. Show that the solution is unique.

Hint Use the substitution $Y(t) = e^{at} X(t)$.

A *linear* stochastic differential equation has the general form

$$dX(t) = aX(t)\, dt + bX(t)\, dW(t), \tag{7.30}$$

where a and b are real numbers. In particular, for $a = 0$ and $b = 1$ we obtain the stochastic differential equation $dX(t) = X(t)\, dW(t)$ in Example 7.6. The solution to the initial value problem for any linear stochastic differential equation can be found by exploiting the analogy with ordinary differential equations, as presented in the exercises below.

Exercise 7.18

Suppose that $w(t)$, $t \geq 0$ is a *deterministic* real-valued function of class C^1 such that $w(0) = 0$. Solve the *ordinary* differential equation

$$dx(t) = ax(t)\, dt + bx(t)\, dw(t), \tag{7.31}$$

with initial condition $x(0) = x_0$ to find that

$$x(t) = x_0 e^{at + bw(t)}. \tag{7.32}$$

(We write $dw(t)$ in place of $w'(t)\, dt$ to emphasize the analogy with stochastic differential equations.)

Hint The variables can be separated:

$$\frac{dx(t)}{x(t)} = \big(a + bw'(t)\big)\, dt.$$

By analogy with the deterministic solution (7.32), let us consider a process defined by

$$X(t) = X_0 e^{at+bW(t)} \tag{7.33}$$

for any $t \geq 0$, where $W(t)$ is a Wiener process.

Exercise 7.19

Show that $X(t)$ defined by (7.33) is a solution of the linear stochastic differential equation

$$dX(t) = \left(a + \frac{b^2}{2}\right) X(t)\, dt + bX(t)\, dW(t), \tag{7.34}$$

with initial condition $X(0) = X_0$.

Hint Use the Itô formula with $F(t,x) = e^{at+bx}$.

Exercise 7.20

Show that the linear stochastic differential equation

$$dX(t) = aX(t)\, dt + bX(t)\, dW(t)$$

with initial condition $X(0) = X_0$ has a unique solution given by

$$X(t) = X_0 e^{(a-\frac{b^2}{2})t+bW(t)}.$$

Hint Apply the result of Exercise 7.19 with suitably redefined constants.

Having solved the general linear stochastic differential equation (7.30), let us consider an example of a non-linear stochastic differential equation. Once again, we begin with a deterministic problem.

Exercise 7.21

Suppose that $w(t)$, $t \geq 0$ is a *deterministic* real-valued function of class C^1 such that $w(0) = 0$. Solve the ordinary differential equation

$$dx(t) = \sqrt{1 + x(t)^2}\, dt + \sqrt{1 + x(t)^2}\, dw(t)$$

with initial condition $x(0) = x_0$.

Hint The variables in this differential equation can be separated.

Exercise 7.22

Show that the process defined by

$$X(t) = \sinh(C + t + W(t)),$$

where $W(t)$ is a Wiener process and $C = \sinh^{-1} X_0$, is a solution of the stochastic differential equation

$$dX(t) = \left(\sqrt{1 + X(t)^2} + \frac{1}{2}X(t)\right) dt + \left(\sqrt{1 + X(t)^2}\right) dW(t)$$

with initial condition $X(0) = X_0$.

Hint Use the Itô formula with $F(t, x) = \sinh(t + x)$.

We shall conclude this chapter with an example of a stochastic differential equation which does not satisfy the assumptions of Theorem 7.7. It turns out that the solution may fail to exist for all times $t \geq 0$. This is a familiar phenomenon in ordinary differential equations. However, stochastic differential equations add a new effect, which does not even make sense in the deterministic case: the maximum time of existence of the solution, called the *explosion time* may be a (non-constant) random variable, in fact a stopping time.

Example 7.8

Consider the stochastic differential equation

$$dX(t) = X(t)^3 dt + X(t)^2 dW(t).$$

Then

$$X(t) = \frac{1}{1 - W(t)}$$

is a solution, which can be verified, at least formally, by using the Itô formula with $F(t, x) = \frac{1}{1-x}$. The solution $X(t)$ exists only up to the first hitting time

$$\tau = \inf \{t \geq 0 : W(t) = 1\}$$

This is the explosion time of $X(t)$. Observe that

$$\lim_{t \nearrow \tau} X(t) = \infty.$$

Strictly speaking, the Itô formula stated in Theorem 7.6 does not cover this case, since $F(t, x) = \frac{1}{1-x}$ has a singularity at $x = 1$. Definition 7.9 does not apply either, as it requires the solution $X(t)$ to be defined for all $t \geq 0$. Suitable extensions of the Itô formula and the definition of a solution are required to study stochastic differential equations involving explosions. However, to prevent an explosion of this book, we have to refer the interested reader to a further course in stochastic analysis.

7.6 Solutions

Solution 7.1

Using the first identity in the hint we obtain

$$\begin{aligned}\sum_{j=0}^{n-1} W(t_j^n)\left(W(t_{j+1}^n) - W(t_j^n)\right) &= \frac{1}{2}\sum_{j=0}^{n-1}\left(W(t_{j+1}^n)^2 - W(t_j^n)^2\right) \\ &\quad -\frac{1}{2}\sum_{j=0}^{n-1}\left(W(t_{j+1}^n) - W(t_j^n)\right)^2 \\ &= \frac{1}{2}W(T)^2 - \frac{1}{2}\sum_{j=0}^{n-1}\left(W(t_{j+1}^n) - W(t_j^n)\right)^2.\end{aligned}$$

By Exercise 6.29 the limit is

$$\lim_{n\to\infty}\sum_{j=0}^{n-1} W(t_j^n)\left(W(t_{j+1}^n) - W(t_j^n)\right) = \frac{1}{2}W(T)^2 - \frac{1}{2}T.$$

Similarly, the second identity in the hint enables us to write

$$\begin{aligned}\sum_{j=0}^{n-1} W(t_{j+1}^n)\left(W(t_{j+1}^n) - W(t_j^n)\right) &= \frac{1}{2}\sum_{j=0}^{n-1}\left(W(t_{j+1}^n)^2 - W(t_j^n)^2\right) \\ &\quad +\frac{1}{2}\sum_{j=0}^{n-1}\left(W(t_{j+1}^n) - W(t_j^n)\right)^2 \\ &= \frac{1}{2}W(T)^2 + \frac{1}{2}\sum_{j=0}^{n-1}\left(W(t_{j+1}^n) - W(t_j^n)\right)^2.\end{aligned}$$

It follows that

$$\lim_{n\to\infty}\sum_{j=0}^{n-1} W(t_{j+1}^n)\left(W(t_{j+1}^n) - W(t_j^n)\right) = \frac{1}{2}W(T)^2 + \frac{1}{2}T.$$

Solution 7.2

For any random step processes $f, g \in M^2_{\text{step}}$ there is a partition $0 = t_0 < t_1 < \cdots < t_n$ such that for any $t \geq 0$

$$f(t) = \sum_{j=0}^{n-1} \eta_j 1_{[t_j, t_{j+1})}(t) \quad \text{and} \quad g(t) = \sum_{j=0}^{n-1} \zeta_j 1_{[t_j, t_{j+1})}(t),$$

where η_j and ζ_j are square integrable $\mathcal{F}_{t_j}$-measurable random variables for each $j = 0, 1, \ldots, n-1$. (If the two partitions in the formulae for f and g happen to be different, then it is always possible to find a common refinement of the two partitions.)

As in the proof of Proposition 7.1, we denote the increment $W(t_{j+1})-W(t_j)$ by $\Delta_j W$ and $t_{j+1}-t_j$ by $\Delta_j t$. Then

$$\begin{aligned} I(f)I(g) &= \sum_{j=0}^{n-1}\sum_{k=0}^{n-1} \eta_j \zeta_k \Delta_j W \Delta_k W \\ &= \sum_{j=0}^{n-1} \eta_j \zeta_j \left|\Delta_j W\right|^2 + \sum_{j<k} \eta_j \zeta_k \Delta_j W \Delta_k W + \sum_{j<k} \zeta_j \eta_k \Delta_j W \Delta_k W, \end{aligned}$$

where, by independence,

$$E\left(\eta_j \zeta_j \Delta_j^2\right) = E\left(\eta_j \zeta_j\right) E\left(\Delta_j^2\right) = E\left(\eta_j \zeta_j\right) \Delta_j t$$

and

$$\begin{aligned} E\left(\eta_j \zeta_k \Delta_j W \Delta_k W\right) &= E\left(\eta_j \zeta_k \Delta_j W\right) E\left(\Delta_k W\right) = 0 \\ E\left(\zeta_j \eta_k \Delta_j W \Delta_k W\right) &= E\left(\zeta_j \eta_k \Delta_j W\right) E\left(\Delta_k W\right) = 0 \end{aligned}$$

for any $j < k$. It follows that

$$E\left(I(f)I(g)\right) = \sum_{j=0}^{n-1} E\left(\eta_j \zeta_j\right) \Delta_j t.$$

Therefore, it suffices to show that

$$E\left(\int_0^\infty f(t)g(t)\,dt\right) = \sum_{j=0}^{n-1} E\left(\eta_j \zeta_j\right) \Delta_j t,$$

but this is true because

$$\begin{aligned} f(t)g(t) &= \sum_{j=0}^{n-1}\sum_{k=0}^{n-1} \eta_j \zeta_k 1_{[t_j,t_{j+1})}(t) 1_{[t_k,t_{k+1})}(t) \\ &= \sum_{j=0}^{n-1} \eta_j \zeta_j 1_{[t_j,t_{j+1})}(t). \end{aligned}$$

Solution 7.3

We shall use a partition $0 = t_0 < t_1 < \cdots < t_n$ such that

$$f = \sum_{j=0}^{n-1} \eta_j 1_{[t_j,t_{j+1})} \quad \text{and} \quad g = \sum_{j=0}^{n-1} \zeta_j 1_{[t_j,t_{j+1})},$$

where η_j and ζ_j are square integrable $\mathcal{F}_{t_j}$-measurable random variables for each $j = 0, 1, \ldots, n-1$. (If the two partitions in the formulae for f and g happen to be different, then it is always possible to find a common refinement of the two partitions.) The increments $W(t_{j+1}) - W(t_j)$ will be denoted by $\Delta_j W$ for brevity. Then

$$\alpha f + \beta g = \sum_{j=0}^{n-1} (\alpha\eta_j + \beta\zeta_j)\, 1_{[t_j,t_{j+1})}$$

and

$$\begin{aligned} I(\alpha f + \beta g) &= \sum_{j=0}^{n-1} (\alpha\eta_j + \beta\zeta_j)\, \Delta_j W \\ &= \alpha \sum_{j=0}^{n-1} \eta_j \Delta_j W + \beta \sum_{j=0}^{n-1} \zeta_j \Delta_j W \\ &= \alpha I(f) + \beta I(g). \end{aligned}$$

Solution 7.4

Consider the following scalar products in M^2 and L^2:

$$\langle f, g\rangle_{M^2} = E\left(\int_0^\infty f(t)g(t)\,dt\right) \quad \text{and} \quad \langle \eta, \zeta\rangle_{L^2} = E(\eta\zeta)$$

for any $f, g \in M^2$ and $\eta, \zeta \in L^2$. They can be expressed in terms of the corresponding norms defined in the proof of Proposition 7.2,

$$\begin{aligned} \langle f, g\rangle_{M^2} &= \frac{1}{4}\|f+g\|_{M^2}^2 - \frac{1}{4}\|f-g\|_{M^2}^2, \\ \langle \eta, \zeta\rangle_{L^2} &= \frac{1}{4}\|\eta+\zeta\|_{L^2}^2 - \frac{1}{4}\|\eta-\zeta\|_{L^2}^2. \end{aligned}$$

Therefore Proposition 7.2 implies that

$$\langle I(f), I(g)\rangle_{L^2} = \langle f, g\rangle_{M^2},$$

which is the same as the equality to be proved.

Solution 7.5

If $f \in M^2_{\text{step}}$ is a random step process, then so is $1_{[0,t]} f \in M^2_{\text{step}} \subset M^2$ for any $t > 0$. This in turn implies that $f \in M^2_t$ for any $t > 0$.

We shall verify that $I_t(f)$ is a martingale with respect to the filtration $\mathcal{F}_t$. Let $0 \le s < t$ and suppose that $f \in M^2_{\text{step}}$ can be written in the form (7.2), where

$$0 = t_0 < t_1 < \cdots < t_k = s < t_{k+1} < \cdots < t_m = t < t_{m+1} < \cdots < t_n.$$

Such a partition $t_0, \ldots, t_n$ can always be obtained by adding the points s and t if necessary. We shall denote the increment $W(t_{j+1}) - W(t_j)$ by $\Delta_j W$ as in the proof of Proposition 7.1. Then

$$1_{[0,t]} f = \sum_{j=0}^{m-1} \eta_j 1_{[t_j, t_{j+1}]}$$

and

$$I_t(f) = I(1_{[0,t]} f) = \sum_{j=0}^{m-1} \eta_j \Delta_j W,$$

which is adapted to $\mathcal{F}_t$ and square integrable, and so integrable. It remains to compute

$$E\left(I_t(f)|\mathcal{F}_s\right) = E\left(I(1_{[0,t]} f)|\mathcal{F}_s\right) = \sum_{j=0}^{m-1} E\left(\eta_j \Delta_j W|\mathcal{F}_s\right).$$

If $j < k$, then η_j and $\Delta_j W$ are $\mathcal{F}_s$-measurable and

$$E\left(\eta_j \Delta_j W|\mathcal{F}_s\right) = \eta_j \Delta_j W.$$

If $j \geq k$, then $\mathcal{F}_s \subset \mathcal{F}_{t_j}$ and

$$\begin{aligned} E\left(\eta_j \Delta_j W|\mathcal{F}_s\right) &= E\left(E\left(\eta_j \Delta_j W|\mathcal{F}_{t_j}\right)|\mathcal{F}_s\right) \\ &= E\left(\eta_j E\left(\Delta_j W|\mathcal{F}_{t_j}\right)|\mathcal{F}_s\right) \\ &= E\left(\eta_j|\mathcal{F}_s\right) E\left(\Delta_j W\right) = 0, \end{aligned}$$

since η_j is $\mathcal{F}_{t_j}$-measurable and $\Delta_j W$ is independent of $\mathcal{F}_{t_j}$. It follows that

$$E\left(I_t(f)|\mathcal{F}_s\right) = \sum_{j=0}^{k-1} \eta_j \Delta_j W = I(1_{[0,s]} f) = I_s(f).$$

Solution 7.6

By definition, $W(t)$ is adapted to the filtration $\mathcal{F}_t$ and has a.s. continuous paths. Moreover,

$$\begin{aligned} E\left(\int_0^T |W(t)|^2 \, dt\right) &= \int_0^T E\left(|W(t)|^2\right) dt \\ &= \int_0^T t \, dt < \infty. \end{aligned}$$

By Theorem 7.1 it follows that the Wiener process W belongs to M_T^2.

Solution 7.7

Since $W(t)$ is adapted to the filtration $\mathcal{F}_t$, so is $W(t)^2$. Moreover,

$$E\left(\int_0^T |W(t)|^4\,dt\right) = \int_0^T E\left(|W(t)|^4\right) dt$$
$$= \int_0^T 3t^2\,dt < \infty.$$

Theorem 7.1 implies that $W(t)^2$ belongs to M_T^2.

Solution 7.8

We fix $T > 0$ and put

$$f(t) = 1_{[0,T)}(t)W(t).$$

Then $f \in M^2$ and

$$\int_0^T W(t)\,dW(t) = \int_0^\infty f(t)\,dW(t).$$

Take $0 = t_0^n < t_1^n < \cdots < t_n^n = T$, where $t_i^n = \frac{iT}{n}$, to be a partition of $[0,T]$ into n equal parts, and put

$$f_n = \sum_{i=0}^{n-1} W(t_i^n) 1_{[t_i^n, t_{i+1}^n)}.$$

Then the sequence $f_1, f_2, \ldots \in M_{\text{step}}^2$ approximates f, since

$$E\left(\int_0^\infty |f(t) - f_n(t)|^2\,dt\right) = \sum_{i=0}^{n-1} \int_{t_i^n}^{t_{i+1}^n} E\left(|W(t) - W(t_i^n)|^2\right) dt$$
$$= \sum_{i=0}^{n-1} \int_{t_i^n}^{t_{i+1}^n} (t - t_i^n)\,dt$$
$$= \frac{1}{2}\sum_{i=0}^{n-1} \left(t_{i+1}^n - t_i^n\right)^2$$
$$= \frac{1}{2}\frac{T^2}{n} \to 0 \quad \text{as } n \to \infty.$$

By Exercise 7.1

$$I(f_n) = \sum_{i=0}^{n-1} W(t_i^n)\left(W(t_{i+1}^n) - W(t_i^n)\right) \to \frac{1}{2}W(T)^2 - \frac{1}{2}T$$

in L^2 as $n \to \infty$. We have found, therefore, that

$$\int_0^T W(t)\,dW(t) = \frac{1}{2}W(T)^2 - \frac{1}{2}T.$$

Solution 7.9

Let $f(t) = t$. Then $1_{[0,T]}f$ belongs to M_T^2. We shall use the same partition of $[0,T]$ into n equal parts as in Solution 7.8. The sequence

$$f_n = \sum_{i=0}^{n-1} t_i^n 1_{[t_i^n, t_{i+1}^n)} \in M^2_{\text{step}}$$

approximates $1_{[0,T]}f$, since

$$\begin{aligned} E\left(\int_0^\infty \left|1_{[0,T]}f(t) - f_n(t)\right|^2 dt\right) &= E\left(\int_0^T |f(t) - f_n(t)|^2\,dt\right) \\ &= \sum_{i=1}^{n-1} \int_{t_i^n}^{t_{i+1}^n} |t - t_i^n|^2\,dt \\ &= \frac{1}{3}\sum_{i=1}^{n-1} \frac{T^3}{n^3} \\ &= \frac{T^3}{3n^2} \to 0 \quad \text{as } n \to \infty. \end{aligned}$$

With the aid of the identity in the hint, we can write the stochastic integral of f_n as

$$\begin{aligned} I(f_n) &= \sum_{i=0}^{n-1} t_i^n \left(W(t_{i+1}^n) - W(t_i^n)\right) \\ &= \sum_{i=0}^{n-1} \left(t_{i+1}^n W(t_{i+1}^n) - t_i^n W(t_i^n)\right) - \sum_{i=0}^{n-1} W(t_{i+1}^n)\left(t_{i+1}^n - t_i^n\right) \\ &= TW(T) - \sum_{i=0}^{n-1} W(t_{i+1}^n)\left(t_{i+1}^n - t_i^n\right). \end{aligned}$$

It follows that

$$I(f_n) \to TW(T) - \int_0^T W(t)\,dt$$

in L^2 as $n \to \infty$, since by the Cauchy-Schwartz inequality

$$E\left(\left|\sum_{i=0}^{n-1} W(t_{i+1}^n)\left(t_{i+1}^n - t_i^n\right) - \int_0^T W(t)\,dt\right|^2\right)$$

$$
\begin{aligned}
&= E\left(\left|\sum_{i=0}^{n-1}\left(\int_{t_i^n}^{t_{i+1}^n}\left(W(t_{i+1}^n)-W(t)\right)dt\right)\right|^2\right)\\
&\le n\sum_{i=0}^{n-1}E\left(\left|\int_{t_i^n}^{t_{i+1}^n}\left(W(t_{i+1}^n)-W(t)\right)dt\right|^2\right)\\
&\le n\sum_{i=0}^{n-1}\left(t_{i+1}^n-t_i^n\right)E\left(\int_{t_i^n}^{t_{i+1}^n}\left|W(t_{i+1}^n)-W(t)\right|^2dt\right)\\
&= n\sum_{i=0}^{n-1}\frac{\left(t_{i+1}^n-t_i^n\right)^3}{2}=n\sum_{i=0}^{n-1}\frac{T^3}{2n^3}=\frac{T^3}{2n}\to 0 \quad \text{as } n\to\infty.
\end{aligned}
$$

This proves the equality in the exercise.

Solution 7.10

Using the same partition of $[0,T]$ into n equal parts as in Solution 7.8 and putting

$$
f_n=\sum_{i=0}^{n-1}W(t_i^n)^2 1_{[t_i^n,t_{i+1}^n)},
$$

we obtain a sequence $f_1,f_2,\ldots\in M^2_{\mathrm{step}}$ of random step processes which approximates $f=1_{[0,T]}W^2$. Indeed,

$$
\begin{aligned}
E\left(\int_0^\infty |f(t)-f_n(t)|^2\,dt\right) &= \sum_{i=0}^{n-1}\int_{t_i^n}^{t_{i+1}^n}E\left(\left|W(t)^2-W(t_i^n)^2\right|^2\right)dt\\
&= \sum_{i=0}^{n-1}\int_{t_i^n}^{t_{i+1}^n}\left(3\left(t-t_i^n\right)^2+4(t-t_i^n)t_i^n\right)dt\\
&= \sum_{i=0}^{n-1}\left[\left(\frac{T}{n}\right)^3+2\left(\frac{T}{n}\right)^2\frac{iT}{n}\right]\\
&= \frac{T^3}{n}\to 0 \quad \text{as } n\to\infty.
\end{aligned}
$$

The expectation above is computed with the aid of the following formula valid for any $0\le s\le t$:

$$
\begin{aligned}
E\left(\left(W_t^2-W_s^2\right)^2\right) &= E\left((W_t-W_s)^4\right)+4E\left((W_t-W_s)^3W_s\right)\\
&\quad +4\left(E\left(W_t-W_s\right)^2W_s^2\right)\\
&= 3\left(t-s\right)^2+4(t-s)s
\end{aligned}
$$

Using the identity in the hint, we can write

$$
\begin{aligned}
I(f_n) &= \sum_{i=0}^{n-1} W(t_i^n)^2 \left(W(t_{i+1}^n) - W(t_i^n)\right) \\
&= \frac{1}{3}\sum_{i=0}^{n-1} \left(W(t_{i+1}^n)^3 - W(t_i^n)^3\right) \\
&\quad - \sum_{i=0}^{n-1} W(t_i^n) \left(W(t_{i+1}^n) - W(t_i^n)\right)^2 - \frac{1}{3}\sum_{i=0}^{n-1} \left(W(t_{i+1}^n) - W(t_i^n)\right)^3 \\
&= \frac{1}{3}W(T)^3 - \sum_{i=0}^{n-1} W(t_i^n) \left(t_{i+1}^n - t_i^n\right) \\
&\quad - \sum_{i=0}^{n-1} W(t_i^n) \left[\left(W(t_{i+1}^n) - W(t_i^n)\right)^2 - \left(t_{i+1}^n - t_i^n\right)\right] \\
&\quad - \frac{1}{3}\sum_{i=0}^{n-1} \left(W(t_{i+1}^n) - W(t_i^n)\right)^3 .
\end{aligned}
$$

The L^2 limits of the last three sums are

$$
\begin{aligned}
\lim_{n\to\infty} \sum_{i=0}^{n-1} W(t_i^n) \left(t_{i+1}^n - t_i^n\right) &= \int_0^T W(t)\, dt \\
\lim_{n\to\infty} \sum_{i=0}^{n-1} W(t_i^n) \left[\left(W(t_{i+1}^n) - W(t_i^n)\right)^2 - \left(t_{i+1}^n - t_i^n\right)\right] &= 0 \\
\lim_{n\to\infty} \sum_{i=0}^{n-1} \left(W(t_{i+1}^n) - W(t_i^n)\right)^3 &= 0
\end{aligned}
$$

Indeed, the first limit is correct because

$$
\begin{aligned}
& E\left(\left|\sum_{i=0}^{n-1} W(t_i^n) \left(t_{i+1}^n - t_i^n\right) - \int_0^T W(t)\, dt\right|^2\right) \\
&= E\left(\left|\sum_{i=0}^{n-1} \int_{t_i^n}^{t_{i+1}^n} \left(W(t_i^n) - W(t)\right) dt\right|^2\right) \\
&= \sum_{i=0}^{n-1} \int_{t_i^n}^{t_{i+1}^n} E\left(\left|W(t_i^n) - W(t)\right|^2\right) dt \\
&= \sum_{i=0}^{n-1} \int_{t_i^n}^{t_{i+1}^n} \left(t - t_i^n\right) dt \\
&= \frac{T^2}{2n} \to 0 \quad \text{as } n \to \infty.
\end{aligned}
$$

To the second limit can be verified as follows:

$$
\begin{aligned}
&E\left(\left|\sum_{i=0}^{n-1} W(t_i^n)\left[\left(W(t_{i+1}^n)-W(t_i^n)\right)^2-\left(t_{i+1}^n-t_i^n\right)\right]\right|^2\right)\\
&=\sum_{i=0}^{n-1} E\left(W(t_i^n)^2\left|\left(W(t_{i+1}^n)-W(t_i^n)\right)^2-\left(t_{i+1}^n-t_i^n\right)\right|^2\right)\\
&=\sum_{i=0}^{n-1} E\left(W(t_i^n)^2\right) E\left(\left|\left(W(t_{i+1}^n)-W(t_i^n)\right)^2-\left(t_{i+1}^n-t_i^n\right)\right|^2\right)\\
&=2\sum_{i=0}^{n-1} t_i^n\left(t_{i+1}^n-t_i^n\right)^2\\
&=\frac{(n-1)}{n^2}T^2\to 0 \quad \text{as } n\to\infty.
\end{aligned}
$$

Finally, for the third limit we have

$$
\begin{aligned}
&E\left(\left|\sum_{i=0}^{n-1}\left(W(t_{i+1}^n)-W(t_i^n)\right)^3\right|^2\right)\\
&=\sum_{i=0}^{n-1} E\left(\left(W(t_{i+1}^n)-W(t_i^n)\right)^6\right)\\
&=6\sum_{i=0}^{n-1}\left(t_{i+1}^n-t_i^n\right)^3\\
&=6\sum_{i=0}^{n-1}\frac{T^3}{n^3}=\frac{6T^3}{n^2}\to 0 \quad \text{as } n\to\infty.
\end{aligned}
$$

It follows that

$$
I(f_n)\to\frac{1}{3}W(T)^3-\int_0^T W(t)\,dt,
$$

which proves the formula in the exercise.

Solution 7.11

We shall use part 2) of Theorem 7.1 to verify that

$$
\xi(t)=\int_0^t W(s)\,dW(s)
$$

belongs to M_T^2 for any $T\geq 0$. By Theorem 7.4 $\xi(t)$ can be identified with an adapted modification having a.s. continuous trajectories. It suffices to verify that $\xi(t)$ satisfies condition (7.9). Since the stochastic integral is an isometry,

$$E\left|\int_0^t W(s)\,dW(s)\right|^2 = E\int_0^t |W(s)|^2\,ds = \int_0^t s\,ds = \frac{t^2}{2}.$$

It follows that

$$E\int_0^T |\xi(t)|^2\,dt = E\int_0^T \left|\int_0^t W(s)\,dW(s)\right|^2 dt = \int_0^T \frac{t^2}{2}\,dt = \frac{T^3}{6} < \infty,$$

i.e. $\xi(t)$ satisfies (7.9). As a result, $\xi(t)$ belongs to $\mathcal{M}_T^2$.

Solution 7.12

We shall use the equality proved in Exercise 7.10:

$$W(T)^3 = 3\int_0^T W(t)\,dt + 3\int_0^T W(t)^2 dW(t).$$

The process $3W(t)$ belongs to $\mathcal{L}_T^1$ for any $T \geq 0$ because it is adapted and has a.s. continuous paths, so the integral $\int_0^T |3W(t)|\,dt$ exists and is finite. By Exercise 7.7 the process $3W(t)^2$ belongs to $\mathcal{M}_T^2$ for any $T \geq 0$. It follows that $W(t)^3$ is an Itô process. Moreover, the above equation can be written in differential form as

$$dW(t)^3 = 3W(t)\,dt + 3W(t)^2\,dW(t),$$

which gives a formula for the stochastic differential $dW(t)^3$.

Solution 7.13

It has been shown in Exercise 7.9 that

$$TW(T) = \int_0^T W(t)\,dt + \int_0^T t\,dW(t).$$

Since the Wiener process $W(t)$ is adapted and has continuous paths, it belongs to $\mathcal{L}_T^1$, while the deterministic process $f(t) = t$ belongs to $\mathcal{M}_T^2$ for any $T > 0$. It follows that $tW(t)$ is an Itô process with stochastic differential

$$d\,(tW(t)) = W(t)\,dt + t\,dW(t).$$

Solution 7.14

For $F(t,x) = e^x e^{-\frac{t}{2}}$ the partial derivatives are $F_t'(t,x) = -\frac{1}{2}e^x e^{-\frac{t}{2}}$, $F_x'(t,x) =$

$e^x e^{-\frac{t}{2}}$ and $F''_{xx}(t,x) = e^x e^{-\frac{t}{2}}$. Since $X(t) = e^{W(t)}e^{-\frac{t}{2}}$, the Itô formula implies that

$$\begin{aligned} dX(t) &= dF(t, W(t)) \\ &= \left(F'_t(t, W(t)) + \frac{1}{2}F''_{xx}(t, W(t))\right) dt + F'_x(t, W(t))\, dW(t) \\ &= \left(-\frac{1}{2}X(t) + \frac{1}{2}X(t)\right) dt + X(t)\, dW(t) \\ &= X(t)\, dW(t). \end{aligned}$$

Because of this, to show that $X(t)$ is an Itô process we need to verify that $X(t) = e^{W(t)}e^{-\frac{t}{2}}$ belongs to M_T^2 for any $T > 0$. Clearly, it is an adapted process. It was computed in Solution 6.35 that $Ee^{W(t)} = e^{\frac{t}{2}}$, so

$$E\int_0^T |X(t)|\, dt = \int_0^T Ee^{W(t)}e^{-\frac{t}{2}} dt = \int_0^T dt = T < \infty,$$

which proves that $X(t)$ belongs to M_T^2.

Solution 7.15

Take $F(t,x) = x^n$. Then $F'_t(t,x) = 0$, $F'_x(t,x) = nx^{n-1}$ and $F''_{xx}(t,x) = n(n-1)x^{n-2}$. The derivatives of $F(t,x)$ are obviously continuous, so we only need to verify that $F(t, W(t)) = nW(t)^{n-1}$ belongs to M_T^2 for $T \geq 0$. Clearly, it is adapted and has a.s. continuous paths. Moreover,

$$E\int_0^T \left|nW(t)^{n-1}\right|^2 dt = n^2 \int_0^T E\left|W(t)\right|^{2n-2} dt = \int_0^T a_{2n-2}t^{n-1}\, dt < \infty,$$

where $a_k = 2^{k/2}\pi^{-1/2}\Gamma(\frac{k+1}{2})$ and $\Gamma(x) = \int_0^\infty t^{x-1}e^{-t}\, dt$ is the Euler gamma function. It follows by part 2) of Theorem 7.1 that $F(t, W(t)) = nW(t)^{n-1}$ belongs to M_T^2. Therefore we can apply the Itô formula to get

$$d\left(W(t)^n\right) = \frac{n(n-1)}{2}W(t)^{n-2}dt + nW(t)^{n-1}dW(t),$$

as required.

Solution 7.16

Some elementary calculus shows that $F(t,x) = e^{-\alpha t}x$ has continuous partial derivatives such that $F'_t(t,x) = -\alpha e^{-\alpha t}x$, $F'_x(t,x) = e^{-\alpha t}$ and $F''_{xx}(t,x) = 0$. Clearly, $\xi(t) = \sigma\int_0^t e^{\alpha s}\, dW(s)$ is an Itô process with

$$d\xi(t) = \sigma e^{\alpha t} dW(t).$$

Since the function $\sigma\varepsilon^{\alpha t}F'_x(t,x)$ is bounded on each set of the form $[0,T]\times\mathbb{R}$, it follows immediately that $\sigma\varepsilon^{\alpha t}F'_x(t,\xi(t))$ belongs to M_T^2 for any $T\geq 0$. As a consequence, we can use the Itô formula (the general case in Theorem 7.6) to get

$$\begin{aligned} dY(t) &= d\left(e^{-\alpha t}\xi(t)\right) \\ &= -\alpha e^{-\alpha t}\xi(t)\,dt + e^{-\alpha t}\sigma e^{\alpha t}\,dW(t) \\ &= -\alpha Y(t)\,dt + \sigma\,dW(t), \end{aligned}$$

which proves that $Y(t)$ satisfies the equality

$$dY(t) = -\alpha Y(t)\,dt + \sigma\,dW(t).$$

Solution 7.17

Take $F(t,x) = e^{\alpha t}x$ and consider the process

$$Y(t) = F(t, X(t)).$$

Then $Y(0) = x_0$ and

$$\begin{aligned} dY(t) &= dF(t,X(t)) \\ &= \left(F'_t(t,X(t)) - \alpha X(t)F'_x(t,X(t)) + \frac{1}{2}\sigma^2 F''_{xx}(t,X(t))\right)dt \\ &\quad + \sigma F'_x(t,X(t))\,dW(t) \\ &= \left(\alpha e^{\alpha t}X(t) - \alpha e^{\alpha t}X(t)\right)dt + \sigma e^{\alpha t}\,dW(t) \\ &= \sigma e^{\alpha t}\,dW(t). \end{aligned}$$

by the Itô formula. It follows that

$$Y(t) = x_0 + \sigma\int_0^t e^{\alpha s}\,dW(s)$$

and

$$\begin{aligned} X(t) &= e^{-\alpha t}Y(t) \\ &= e^{-\alpha t}x_0 + \sigma e^{-\alpha t}\int_0^t e^{\alpha s}\,dW(s). \end{aligned}$$

Uniqueness follows directly from the above argument, but Theorem 7.7 can also be used. Namely, the stochastic differential equation

$$dX(t) = -\alpha X(t)\,dt + \sigma\,dW(t)$$

is of the form (7.24) with $f(x) = -\alpha x$ and $g(x) = \sigma$, which are Lipschitz continuous functions. Therefore, the solution to the initial value problem must be unique in the class of Itô processes with a.s. continuous paths.

Solution 7.18

According to the theory of ordinary differential equations, (7.31) with initial condition $x(0) = x_0$ has a unique solution. If $x_0 = 0$, then $x(t) = 0$ is the solution. If $x_0 \neq 0$, then

$$\ln \frac{x(t)}{x_0} = at + w(t)$$

by integrating the equation in the hint, which implies that

$$x(t) = x_0 e^{at+bw(t)}.$$

Solution 7.19

By the Itô formula (verify the assumptions!)

$$\begin{aligned} dX(t) &= d\left(X_0 e^{at+bW(t)}\right) \\ &= \left(aX_0 e^{at+bW(t)} + \frac{b^2}{2} X_0 e^{at+bW(t)}\right) dt + bX_0 e^{at+bW(t)} dW(t) \\ &= \left(a + \frac{b^2}{2}\right) X(t)\, dt + bX(t)\, dW(t). \end{aligned}$$

This proves that $X(t)$ satisfies the stochastic differential equation (7.34). As regards the initial condition, we have

$$X(0) = X_0 e^{at+bW(t)}\Big|_{t=0} = X_0.$$

Solution 7.20

The stochastic differential equation

$$dX(t) = aX(t)\, dt + bX(t)\, dW(t)$$

can be written as

$$dX(t) = \left(c + \frac{b^2}{2}\right) X(t)\, dt + bX(t)\, dW(t),$$

where $c = a - \frac{b^2}{2}$. By Exercise 7.19 the solution this stochastic differential equation with initial condition $X(0) = X_0$ is

$$\begin{aligned} X(t) &= X_0 e^{ct+bW(t)} \\ &= X_0 e^{\left(a-\frac{b^2}{2}\right)t+bW(t)}. \end{aligned}$$

The uniqueness of this solution follows immediately from Theorem 7.7.

Solution 7.21

We can write the ordinary differential equation to be solved in the form

$$\frac{dx(t)}{\sqrt{1+x(t)^2}} = (1+w'(t))\,dt,$$

which implies that

$$\sinh^{-1} x(t) - \sinh^{-1} x_0 = t + w(t).$$

Composing the last formula with sinh, we obtain

$$x(t) = \sinh(c + t + w(t)), \tag{7.35}$$

where $c = \sinh^{-1} x_0$.

Solution 7.22

Since $F(t,x) = \sinh(t+x)$ satisfies the assumptions of the Itô formula,

$$\begin{aligned}
dX(t) &= dF(t, C + W(t)) \\
&= \left(F'_t(t, C+W(t)) + \frac{1}{2}F''_{xx}(t, C+W(t))\right) dt \\
&\quad + F'_x(t, C+W(t))\,dW(t) \\
&= \left(\cosh(C+t+W(t)) + \frac{1}{2}\sinh(C+t+W(t))\right) dt \\
&\quad + \cosh(C+t+W(t))\,dW(t) \\
&= \left(\sqrt{1+\sinh^2(C+t+W(t))} + \frac{1}{2}\sinh(C+t+W(t))\right) dt \\
&\quad + \sqrt{1+\sinh^2(C+t+W(t))}\,dW(t) \\
&= \left(\sqrt{1+X(t)^2} + \frac{1}{2}X(t)\right) dt + \left(\sqrt{1+X(t)^2}\right) dW(t).
\end{aligned}$$

The initial condition $X(0) = \sinh C = X_0$ is also satisfied.

Index